Machine Learning Applications

Machine Learning Applications

From Computer Vision to Robotics

Edited by

Indranath Chatterjee
Department of Computer Engineering
Tongmyong University
Busan, South Korea

Sheetal Zalte
Department of Computer Science
Shivaji University
Kolhapur, Maharashtra, India

IEEE PRESS

WILEY

Published by John Wiley & Sons, Inc., Hoboken, New Jersey.
Published simultaneously in Canada.

For general information on our other products and services or for technical support, please contact our Customer Care Department within the United States at (800) 762-2974, outside the United States at (317) 572-3993 or fax (317) 572-4002.

Wiley also publishes its books in a variety of electronic formats. Some content that appears in print may not be available in electronic formats. For more information about Wiley products, visit our web site at www.wiley.com.

Library of Congress Cataloging-in-Publication Data

Names: Chatterjee, Indranath, editor. | Zalte, Sheetal S., editor. | John
    Wiley & Sons, publisher.
Title: Machine learning applications : from computer vision to robotics /
    edited by Indranath Chatterjee, Sheetal Zalte.
Description: Hoboken, New Jersey : Wiley-IEEE Press, [2024] | Includes
    bibliographical references and index.
Identifiers: LCCN 2023043411 (print) | LCCN 2023043412 (ebook) | ISBN
    9781394173327 (cloth) | ISBN 9781394173334 (adobe pdf) | ISBN
    9781394173341 (epub)
Subjects: LCSH: Machine learning–Industrial applications. | Machine
    learning–Scientific applications. | Deep learning (Machine
    learning)–Industrial applications. | Deep learning (Machine
    learning)–Scientific applications.
Classification: LCC Q325.5 .M321323 2024 (print) | LCC Q325.5 (ebook) |
    DDC 006.3/1–dc23/eng/20231023
LC record available at https://lccn.loc.gov/2023043411
LC ebook record available at https://lccn.loc.gov/2023043412

Cover Design: Wiley
Cover Image: © Yuichiro Chino/Getty Images

Set in 9.5/12.5pt STIXTwoText by Straive, Pondicherry, India

Contents

About the Authors

Dr. Indranath Chatterjee is working as a professor in the Department of Computer Engineering at Tongmyong University, Busan, South Korea. He received his PhD in Computational Neuroscience from the Department of Computer Science, University of Delhi, Delhi, India. His research areas include computational neuroscience, schizophrenia, medical imaging, fMRI, and machine learning. He has authored and edited nine books on computer science and neuroscience published by renowned international publishers. To date, he has published more than 70 research papers in international journals and conferences. He is the recipient of various global awards in neuroscience. He is serving as the Chief Section Editor of a few renowned international journals and as a member of the advisory board and editorial board of various international journals and open-science organizations worldwide. He is working on several projects of government and non-government organizations as PI/co-PI, related to medical imaging and machine learning for a broader societal impact, in collaboration with several universities globally. He is an active professional member of the Association of Computing Machinery (ACM, USA), the Organization of Human Brain Mapping (OHBM, USA), the Federation of European Neuroscience Society (FENS, Belgium), the Association for Clinical Neurology and Mental Health (ACNM, India), and the International Neuroinformatics Coordinating Facility (INCF, Sweden).

Dr. Sheetal S. Zalte-Gaikwad is an assistant professor in the Computer Science Department at Shivaji University, Kolhapur, India. She pursued BSc and MSc from Pune University, India. She earned her PhD in mobile ad-hoc network at Shivaji University, India. She has 14 years of teaching experience in computer science. She has published more than 40 research papers in reputed international journals and conferences. She has also published book chapters with Springer, Bentham, CRC Taylor, Wiley, and Francis. Her research areas are MANET, VANET, blockchain security. She has also authored a book, *Computational Theory, Problems and Solutions*. She worked as the lead editor for the book, *Synergistic Interaction of Big Data with Cloud Computing for Industry 4.0*, CRC Press, Taylor and Francis Publisher, USA.

Preface

In our rapidly evolving world, the transformative power of machine learning (ML) and deep learning (DL) technologies is undeniable. From robotics and vehicle automation to financial services, retail, manufacturing, healthcare, and beyond, ML and DL are revolutionizing industries and driving improvements in business operations. The potential of these advanced technologies to enhance our lives and reshape our future is immense.

In this book, we delve into the remarkable advancements made possible by ML and DL, showcasing case studies that demonstrate how these technologies have facilitated breakthroughs in business intelligence, enabling faster and more efficient decision-making processes. We explore a wide range of applications, from facial recognition to natural language processing, and illustrate how ML and DL play a central role in the continuous learning and data simulation capabilities of cars in real-time.

While it is crucial to acknowledge the potential challenges and implications associated with ML and DL, it is equally important to recognize the positive impact they can have on our society. This book aims to shed light on real-world examples that highlight how ML and DL can create better technology to support modern thinking. Whether you are a novice or a specialist in the field, these captivating case studies will offer valuable insights into various applications where ML and DL techniques play a significant role.

Within these pages, we uncover the inner workings of ML algorithms, revealing how they transform digital images, which are mere series of numbers, into meaningful patterns through image processing techniques. We also explore the complex landscapes of risk modeling, genomic sequencing, and modeling, where ML and DL implementations require extensive cloud environments with high-performance data processing and management capabilities.

Moreover, we examine the competitive landscape of ML- and DL-based platforms, where major vendors such as Amazon, Google, Microsoft, IBM, and others vie for customers by offering comprehensive services encompassing data collection, classification, modeling, training, and application deployment.

The revolutionizing influence of ML and DL technologies transcends boundaries, revolutionizing nearly every industry worldwide. This book is dedicated to providing extensive coverage of these groundbreaking technologies and illustrating how they are reshaping industries and our lives.

We explore the vast domain of computer vision and its wide-ranging applications, from everyday life scenarios to the Internet of Things and brain–computer interfaces. With the ability to detect and track humans across multiple streams of data, ML and computer vision represent significant leaps forward, offering tremendous potential in terms of efficiency, productivity, revenue generation, and profitability.

We also examine the critical role played by ML and computer vision in our digital society. They empower individuals with great ideas and limited resources to succeed in business while also enabling established enterprises to harness and analyze the data they collect. Moreover, we highlight how ML contributes to cybersecurity by effectively tracking and preventing monetary frauds online, using examples like PayPal's ML-powered tools for detecting money laundering.

Throughout this book, we aim to cultivate an understanding of the vital importance of ML and computer vision in our AI-driven era. By exploring real-world applications across diverse disciplines and daily-life scenarios, we hope to provide readers with state-of-the-art algorithms and practical insights that underscore the value of AI in future applications.

Embark on this journey with us as we uncover the exciting world of ML and DL, where cutting-edge technology meets real-world impact. May this book empower you to grasp the immense potential of these technologies and inspire you to explore and contribute to their further advancement.

Enjoy the exploration!

November 2023

Indranath Chatterjee
Busan, South Korea
Sheetal Zalte
Kolhapur, Maharashtra, India

1

Statistical Similarity in Machine Learning

Dmitriy Klyushin

Department of Computer Science and Cybernetics, Kyiv, Ukraine

1.1 Introduction

In machine learning, the accuracy of algorithms depends on how accurately the hypothesis about the proximity of objects in the feature space is fulfilled. It is this property that guarantees the possibility of generalization based on training samples. The hypothesis of proximity (similarity) of objects in the feature space assumes that objects of one class form a compact region with a smooth boundary. A classic demonstration of this conjecture is the famous Fisher iris problem, in which points of three classes form easily separable and dense clouds on a plane. This problem illustrates both the strength and the weakness of the compactness hypothesis. The strength of this hypothesis is that we can easily draw boundaries between sets of points and classify them. The weakness of the compactness hypothesis is that that we cannot generalize it to the case when the object is defined not by one point, but by many points. Such situations often arise in medical research, when we take a lot of cells from a patient and measure different features of these cells. As a result, a patient is represented not by a vector in a feature space, but by a matrix of feature samples (moreover, the order of the numbers in the columns of this matrix is random). Of course, it is possible to reduce this matrix to a vector by averaging values in the columns and considering only a vector of means, but it is obvious that this leads to losing of important information about the distribution of feature values. In fact, what can be said about a distribution, knowing only the estimate of its mathematical expectation?

Machine Learning Applications: From Computer Vision to Robotics, First Edition.
Edited by Indranath Chatterjee and Sheetal Zalte.

The hypothesis of compactness ignores the randomness of training data, so we must replace it with an alternative postulate on the proximity between random samples, guided by the laws of mathematical statistics. We propose to use the well-known concept of sample homogeneity in mathematical statistics, i.e. a hypothesis that samples are drawn from a same distribution. Returning to the terminology of machine learning, this means that samples of features of objects have identical distributions. Within this approach, we can use a wide variety of statistical criteria to test the homogeneity hypothesis.

In the chapter, we introduce an alternative concept of proximity in machine learning and propose to use the hypothesis about homogeneity of samples instead of the hypothesis of compactness, as well as provide examples of its effective use.

1.2 Featureless Machine Learning

The pioneers of featureless, or relational machine learning, were scientific schools of Duin (Duin et al. 1997, 1999; Pekalska and Duin 2001; Pekalska and Duin 2005) and Mottl (Mottl et al. 2001, 2002, 2017; Seredin et al. 2012). Their idea was to replace a feature vector of an object by a similarity measure to the training dataset using a metrics. It is obvious that this is not a solution of the problem of classification of objects using a matrix of feature values. The point is that in such cases, it is necessary to use not geometric but statistical tools, for example, two-sample tests of homogeneity, such as the Kolmogorov–Smirnov test and the Mann–Whitney–Wilcoxon test. Using these criteria, we can test the hypothesis that feature samples are homogeneous. However, this is not a complete solution of the posed problem. Testing the homogeneity hypothesis using the tests mentioned above and various other tests, for example, Cramer–von Mises and Anderson–Darling, we cannot obtain a numerical measure of similarity. These tests provide only p-values that denote the probability of the samples being homogeneous. We shall describe a solution allowing measuring the similarity between samples as follows.

To fill the distance matrix, Euclidean and pseudo-Euclidean distances, as well as kernels, are used. It is quite obvious that such an approach is not acceptable for estimating the similarity between matrices whose columns are random samples of features. The use of metrics in such cases is impossible.

Recently, the minimal learning machine (Kuli 2013) and the extreme minimal learning machine (Souza Junior et al. 2015) were developed. The authors used nonlinear distance regression, estimating dissimilarity between objects. There are numerous metrics and learning techniques in this field (Mesquita et al. 2017; Caldas et al. 2018; Florêncio et al. 2018; Maia et al. 2018; Cao et al. 2019; Kärkkäinen 2019;

Bicego 2020; Florêncio et al. 2020; Nanni et al. 2020; Silva et al. 2020 etc.) Details of the surveys of these issues are provided in (Costa et al. 2020; Hämäläinen et al. 2020). All these methods use the Euclidean distance. Therefore, they are unacceptable for solving the problem stated above: to classify objects represented by matrices of independent identically distributed random values.

Our goal is to extend the featureless approach to similarity-based classification using the nonparametric similarity measure and nonparametric two-sample test of homogeneity. Due to the nonparametric nature of these tools, we do not use any assumption about a hypothetical distribution of training sample. Also, as we shall demonstrate below, these tools are universal in the sense that using the proposed test, we can test the homogeneity hypothesis for all possible variants: different location parameters and the same scale parameter, the same location parameter and different scale parameters, and both different location and scale parameters. The proposed similarity measure also is universal because it is applicable to both samples without ties and with ties (duplicates).

1.3 Two-Sample Homogeneity Measure

Consider training samples $a = (a_1, a_2, ..., a_n) \in A$ and $b = (b_1, b_2, ..., b_n) \in B$ from populations A and B obeying distributions F and G that are absolutely continuous. The classification problem for a test sample $c = (c_1, c_2, ..., c_n)$ is reduced to testing the homogeneity of c and a and c and b. There are various nonparametric two-sample tests of homogeneity (Derrick et al. 2019). However, every test has own drawbacks. For example, the Kolmogorov–Smirnov test is a universal test in the sense that it tests the general hypothesis $F = G$, but it is very sensible to outliers and need in large size of samples. The Wilcoxon sign rank test is not universal because it tests only the hypothesis about location shift (i.e. whether $E(a)$ significantly differs from $E(c)$). In our opinion, the most effective and universal tool was developed in Klyushin and Petunin (2003).

1.4 The Klyushin–Petunin Test

The two-sample test of homogeneity (Klyushin and Petunin 2003) is nonparametric. This test uses the Hill's assumption (Hill 1968): for exchangeable random values $a_1, a_2, ..., a_n \in F$ with continuous distribution, we have

$$P\left(x \in \left(a_{(i)}, a_{(j)}\right)\right) = \frac{j-i}{n+1}, j < i, \tag{1.1}$$

where $a_{(i)}$ and $a_{(j)}$ are order statistics and x obeys F.

Let us find $h_{ij} = \dfrac{\#\left\{c_k \in \left(a_{(i)}, a_{(j)}\right)\right\}}{n}$ and estimate the deviation of the observable relative frequency h_{ij} from the expected probability (1.1) constructing a confidence interval for a probability of success in the Bernoulli scheme (Pires and Amado 2008). Since p-statistics is invariant in respect of the selection of a confidence interval for binomial proportion (Klyushin and Martynenko 2021), we may select the most simple one for computations, the Wilson confidence interval $I_{ij}^{(n)} = \left(p_{ij}^{(1)}, p_{ij}^{(2)}\right)$, where

$$
p_{ij}^{(1)} = \frac{h_{ij}n + 0.5g^2 - g\sqrt{h_{ij}\left(1 - h_{ij}\right)n + 0.25g^2}}{n + g^2},
$$

$$
p_{ij}^{(2)} = \frac{h_{ij}n + 0.5g^2 + g\sqrt{h_{ij}\left(1 - h_{ij}\right)n + 0.25g^2}}{n + g^2}.
$$

(1.2)

If $g = 3$, the confidence level of (1.2) is greater than 0.95 (Klyushin and Petunin 2003). Since the number of all the intervals $(a_{(i)}, a_{(j)})$ where $i < j$ is equal to $N = \dfrac{n(n-1)}{2}$, the homogeneity measure for samples a and c is

$$
h = \frac{1}{N} \# \left\{ p_{ij} = \frac{j - i}{n + 1} \in I_{ij}^{(n)} \right\}.
$$

(1.3)

Note that h in (1.3) is also a binomial proportion. Therefore, the test for homogeneity may be formulated in the following way: samples are homogeneous if the confidence interval for the binomial h covers 0.95, else it is rejected.

1.5 Experiments and Applications

Consider the results of two numerical experiments in which samples were drawn from the normal distribution Gaussian(a, 1) and Gaussian(0, 1) and Gaussian(0, $1 - a$) and Gaussian(0, 1). We considered 100 pairs of samples containing 100 random numbers. The p-statistics and p-value of the Kolmogorov–Smirnov statistics (KS-statistics) were averaged. The null hypothesis is accepted if the p-statistic is greater than 0.95 or the p-value of KS-test is less than 0.05. We tested hypotheses about shift of location and scale parameters. In the first case, the null hypothesis supposes that distributions have the same mathematical expectation. In the latter case, the null hypothesis

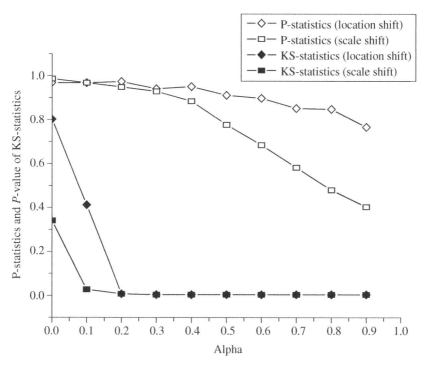

Figure 1.1 Behavior of P-and KS-statistics in the testing the location and scale shift hypothesis.

supposes that distributions have the same standard deviations. The results are presented in Figure 1.1.

In Figure 1.1, we see that the p-statistics decreases as α increases. The point where we can reject the null hypothesis about both location and scale shift is $\alpha = 0.3$. The Kolmogorov–Smirnov test detects the shift of location when $\alpha = 0.1$ and the scale shift when $\alpha = 0.2$.

The graph demonstrates very high sensitivity of both tests. But, the high sensitivity of KS-statistics has a negative side because for $\alpha > 0.2$ the p-value of the KS-test varies close to 0. Therefore, it just recognizes the fact that distributions are different but does not estimate the value of this difference. In contrast, the graph of p-statistics is monotonic and may be used as a similarity measure between statistics in all the ranges of the parameter α. These results may be easily reproduced for pairs of samples drawn from various distributions (lognormal, uniform, gamma et al.). The examples are observed in Klyushin and Martynenko (2021).

The test based on the p-statistics successfully estimated the similarity and difference between the feature samples of patients with breast cancer (Andrushkiw et al. 2007), detected change points in time series (Klyushin and Martynenko 2021), and compared forecasting models for the COVID-19 epidemic curve (Klyushin 2021). The applications of the p-statistics are not bounded by the abovementioned problems. It may be useful to solve the problem if two rankings come from the same distribution (Balázs et al. 2022) and constructing a statistical depth (Goibert et al. 2022), for instance. At the best case, the proposed test is more effective due to its universal nature; in the worst case, it is as effective as the Kolmogorov–Smirnov and other tests (Klyushin 2021). The p-statistics is so-called "soft" similarity measure. In contrast to other tests, the p-statistics is stable with respect to outliers and anomalities. Therefore, it is a natural measure of similarity between two samples.

1.6 Summary

The ability of a machine learning algorithm to generalize results obtained from training datasets depends on the underlying hypotheses. Classical discriminant analysis uses the compactness hypothesis. This geometric hypothesis is not applicable to the classification of random samples since for such samples, the concept of distance is meaningless. We propose an alternative concept of homogeneity of objects, which are considered homogeneous if their sample features have identical distribution. The Klyushin–Petunin two-sample test successfully tests the hypotheses about location and scale shift with high sensitivity and specificity. The future scope of the work is related with the development of analogous tests for multivariate samples.

References

Andrushkiw, R.I., Boroday, N.V., Klyushin, D.A., and Petunin, Y.I. (2007). *Computer-Aided Cytogenetic Method of Cancer Diagnosis*. New York: Nova Publishers.

Balázs, R., Baranyi, M., and Héberger, K. (2022). Testing rankings with cross-validation. *arXiv* https://doi.org/10.48550/arXiv.2105.11939.

Bicego, M. (2020). Dissimilarity random forest clustering. *IEEE International Conference on Data Mining (ICDM)*, Sorrento, Italy (17–20 November 2020), pp. 936–941. IEEE. https://doi.org/10.1109/ICDM50108.2020.00105.

Caldas, W.L., Gomes, J.P.P., and Mesquita, D.P.P. (2018). Fast Co-MLM: an efficient semi-supervised co-training method based on the minimal learning machine. *New Generation Computing* 36: 41–58. https://doi.org/10.1007/s00354-017-0027-x.

Cao, H., Bernard, S., and Sabourin, R.&, Heutte, L. (2019). Random forest dissimilarity based multi-view learning for radiomics application. *Pattern Recognition* 88: 185–197. https://doi.org/10.1016/j.patcog.2018.11.011.

Costa, Y.M.G., Bertolini, D., Britto, A.S. et al. (2020). The dissimilarity approach: a review. *Artificial Intelligence Review* 53: 2783–2808. https://doi.org/10.1007/s10462-019-09746-z.

Derrick, B., White, P., and Toher, D. (2019). Parametric and non-parametric tests for the comparison of two samples which both include paired and unpaired observations. *Journal of Modern Applied Statistical Methods* 18: eP2847. https://doi.org/10.22237/jmasm/1556669520.

Duin, R.P.W., de Ridder, D., and Tax, D.N.J. (1997). Experiments with a featureless approach to pattern recognition. *Pattern Recognition Letters* 18: 1159–1166. https://doi.org/10.1016/S0167-8655(97)00138-4.

Duin, R.P.W., Pekalska, E., and de Ridder, D. (1999). Relational discriminant analysis. *Pattern Recognition Letters* 20: 1175–1181. https://doi.org/10.1016/S0167-8655(99)00085-9.

Florêncio, J.A., Dias, M.L.D., and de Souza Junior, A.H. (2018). A fuzzy c-means-based approach for selecting reference points in minimal learning machines. In: *Fuzzy Information Processing* (ed. G.A. Barreto and R. Coelho), 398–407. Cham: Springer International Publishing. https://doi.org/10.1007/978-3-319-95312-0_34.

Florêncio, J.A., Oliveira, S.A., Gomes, J.P., and da Rocha Neto, A.R. (2020). A new perspective for minimal learning machines: a lightweight approach. *Neurocomputing* 401: https://doi.org/10.1016/j.neucom.2020.03.088.

Goibert, M., Clémençon, S., Irurozki, E., and Mozharovskyi, P. (2022). Statistical depth functions for ranking distributions: definitions, statistical learning and applications. *Proceedings of the 25th International Conference on Artificial Intelligence and Statistics AISTATS 2022*, Valence, Spain (28–30 March 2022). https://hal.archives-ouvertes.fr/hal-03537148/document. https://doi.org/10.48550/arXiv.2201.08105.

Hämäläinen, J., Alencar, A., Kärkkäinen, T. et al. (2020). Minimal learning machine: theoretical results and clustering-based reference point selection. *Journal of Machine Learning Research* 21: 1–29. http://jmlr.org/papers/v21/19-786.html.

Hill, B.M. (1968). Posterior distribution of percentiles: bayes' theorem for sampling from a population. *Journal of the American Statistical Association* 63: 677–691.

Kärkkäinen, T. (2019). Extreme minimal learning machine: ridge regression with distance-based basis. *Neurocomputing* 342: 33–48. https://doi.org/10.1016/j.neucom.2018.12.078.

Klyushin, D. (2021). Non-parametric k-sample tests for comparing forecasting models. *Polibits* 62: 33–41. http://www.polibits.gelbukh.com/2020_62/Non-Parametric%20k-Sample%20Tests%20for%20Comparing%20Forecasting%20Models.pdf. https://doi.org/10.17562/PB-62-4.

Klyushin, D. and Martynenko, I. (2021). Nonparametric test for change point detection in time series. *Proceeding of 3rd International Workshop 'Modern Machine Learning Technologies and Data Science'*, MoMLeT&DS 2021. Volume I: Main Conference, Lviv-Shatsk, Ukraine (5–6 June 2021), pp. 117–127. https://ceur-ws.org/Vol-2917/paper11.pdf (accessed 12 November 2022).

Klyushin, D.A. and Petunin, Y.I. (2003). A nonparametric test for the equivalence of populations based on a measure of proximity of samples. *Ukrainian Mathematical Journal* 55: 181–198. https://doi.org/10.1023/A:1025495727612.

Kulis, B. (2013). Metric learning: a survey. *Foundations and Trends in Machine Learning* 5: 287–364. https://doi.org/10.1561/2200000019.

Maia, A.N., Dias, M.L.D., Gomes, J.P.P., and da Rocha Neto, A.R. (2018). Optimally selected minimal learning machine. In: *Intelligent Data Engineering and Automated Learning – IDEAL* (ed. H. Yin, D. Camacho, P. Novais, and A.J. Tallón-Ballesteros), 670–678. Cham: Springer International Publishing. https://doi.org/10.1007/978-3-030-33617-2.

Mesquita, D.P.P., Gomes, J.P.P., and de Souza Junior, A.H. (2017). Ensemble of efficient minimal learning machines for classification and regression. *Neural Processing Letters* 46: 751–766. https://doi.org/10.1007/s11063-017-9587-5.

Mottl, V., Dvoenko, S., Seredin, O. et al. (2001). Featureless pattern recognition in an imaginary Hilbert space and its application to protein fold classification. *Machine Learning and Data Mining in Pattern Recognition*, Leipzig, Germany (25–27 July 2001), pp. 322–336. Lecture Notes in Computer Science, 2123. https://doi.org/10.1007/3-540-44596-X_26.

Mottl, V., Seredin, O., Dvoenko, S. et al. (2002). Featureless pattern recognition in an imaginary Hilbert space. *International Conference on Pattern Recognition* 2: 88–91. https://doi.org/10.1109/ICPR.2002.1048244.

Mottl, V., Seredin, O., and Krasotkina, O. (2017). Compactness hypothesis, potential functions, and rectifying linear space. *Machine Learning: International Conference Commemorating the 40th Anniversary of Emmanuil Braverman's Decease*, Boston, MA, USA (28–30 April 2017), Invited Talks. https://doi.org/10.1007/978-3-319-99492-5_3.

Nanni, L., Rigo, A., Lumini, A., and Brahnam, S. (2020). Spectrogram classification using dissimilarity space. *Applied Sciences* 10: 4176. https://doi.org/10.3390/app10124176.

Pekalska, E. and Duin, R.P.W. (2001). On combining dissimilarity representations. In: *Multiple Classifier Systems,*. LNCS, 2096 (ed. J. Kittler and F. Roli), 359–368. Berlin: Springer–Verlag. https://doi.org/10.1007/3-540-48219-9_36.

Pekalska, E. and Duin, R.P.W. (2005). *The Dissimilarity Representation for Pattern Recognition, Foundations and Applications*. Singapore: World Scientific.

Pires, A.M. and Amado, C. (2008). Interval estimators for a binomial proportion: Comparison of twenty methods. *REVSTAT–Statistical Journal* 6 (2): 165–197. https://doi.org/10.57805/revstat.v6i2.63.

Seredin O., Mottl, V., Tatarchuk, A. et al. (2012). Convex support and relevance vector machines for selective multimodal pattern recognition. *Proceedings of the 21st International Conference on Pattern Recognition (ICPR2012)*, Tsukuba, Japan (11–15 November 2012), pp. 1647–1650. IEEE.

da Silva, A.C.F., Saïs, F., Waller, E., and Andres, F. (2020). Dissimilarity-based approach for identity link invalidation. *IEEE 29th International Conference on Enabling Technologies: Infrastructure for Collaborative Enterprises (WETICE)*, Bayonne, France (10–13 September 2020), pp. 251–256. IEEE. https://doi. org/10.1109/WETICE49692.2020.00056.

de Souza Junior, A.H., de Corona, F., Barreto, G.A. et al. (2015). Minimal learning machine: a novel supervised distance-based approach for regression and classification. *Neurocomputing* 164: 34–44. https://doi.org/10.1016/ j.neucom.2014.11.073.

2

Development of ML-Based Methodologies for Adaptive Intelligent E-Learning Systems and Time Series Analysis Techniques

Indra Kumari¹, Indranath Chatterjee², and Minho Lee¹

¹ *Korea Institute of Science and Technology Information (KISTI), University of Science and Technology (UST), Daejeon, South Korea*
² *Department of Computer Engineering, Tongmyong University, Busan, South Korea*

2.1 Introduction

Artificial intelligence (AI)'s machine learning (ML) subfield focuses on creating and studying AI software that can teach itself new skills. Definition: ML is the study of how to program computers to learn and make decisions in ways that are indistinguishable from human intelligence (Sarker 2021). The term "machine learning" refers to a technique whereby a computer is taught to optimize a performance metric by analyzing and learning from examples. Generalization and representation are at the heart of ML. The system's ability to generalize to novel data samples is a key feature. According to Herbert Simon, "learning" is the process through which a system experiences adaptive alterations that improve its performance on a given task or collection of activities the next time it is used. If the program's performance on tasks in class T improves with experience E, as measured by the performance measure P, then we say that the program has learned from its past performance and can apply that knowledge to future performance. Tom Mitchell explains that "a computer program is said to learn from experience E concerning some class of tasks T and performance measure P." Robots with AI can learn from their experiences, identify patterns, and infer their meaning (Patel and Patel 2016).

ML and AI have become so pervasive in our daily lives that they are no longer the purview of specialized researchers trying to crack a difficult issue. Instead of being a fluke, this development has a very natural feel to it. Organizations are now able to harness a massive amount of data in developing solutions with far-reaching

Machine Learning Applications: From Computer Vision to Robotics, First Edition.
Edited by Indranath Chatterjee and Sheetal Zalte.
© 2024 The Institute of Electrical and Electronics Engineers, Inc.
Published 2024 by John Wiley & Sons, Inc.

commercial benefits, thanks to the exponential development in processing speed and the introduction of better algorithms for tackling complicated and tough issues. The availability of rich data, new algorithms, and unique methodologies in its numerous applications make financial services, banking, and insurance one of the most important industries with a very high potential in reaping the advantages of ML and AI. Because companies have only scratched the surface of quickly developing fields like deep neural networks and reinforcement learning, the potential of employing these approaches in many applications remains significantly untapped.

Organizations are reaping the benefits of cutting-edge ML applications in areas such as customer segmentation for targeted marketing of newly released products, the development of optimal portfolio strategies, the identification and prevention of money laundering and other illegal activities in the financial markets, the implementation of more intelligent and effective credit risk management practices, and the maintenance of compliance with regulatory frameworks in financial, accounting, and other operational areas. However, the full potential of ML and AI has yet to be discovered or used. Businesses need to take use of these features if they want to gain and keep a competitive edge over the long run. One of the main reasons for the slow adoption of AI/ML models and methods in financial applications is the lack of familiarity and trust in deploying them in critical and privacy-sensitive applications. However, the "black-box" nature of such models and frameworks that analyzes their internal operations in producing outputs and their validations also impedes faster acceptance and deployment of such models in real-world settings.

2.1.1 Machine Learning

All intelligent applications today take advantage of ML's capabilities. By analyzing large amounts of data automatically, ML can help uncover insights that would otherwise be inaccessible through conventional software (Malakar et al. 2019). In recent years, ML has expanded into every area of study. The advent of ML has made it possible for computers to teach themselves. It is implemented to endow machines with the capacity for judgment equal to that of humans. An individual's proficiency with a machine grows with time, and that time is spent accumulating knowledge in the form of data that may be used to teach the machine (Masetic and Subasi 2016).

2.1.2 Types of Machine Learning

There are three main categories of ML, which are as follows:

1) **Supervised learning**: For this method of training, the training data explicitly map outcomes to their corresponding inputs. The system "learns" by parsing the training data for patterns in how inputs lead to desired outcomes.

Supervised learning is useful for two distinct sorts of problems: classification and regression. The goal of classification is to organize data into meaningful groups. Predicting the value of a variable or variables is the goal of regression analysis (Kumar et al. 2021).

2) **Unsupervised learning**: To train a computer in unsupervised learning, it is fed data that has not been labeled in any way, and the program must figure out what those patterns are. Through its algorithms, the system compiles a synthesis of the information. The process of clustering is an example of unsupervised learning. Clustering is assembling things into distinct clusters. When items that share similarities are grouped, we call this a cluster (Qiu et al. 2016).

3) **Reinforcement learning**: The purpose of learning in a reinforcement setting is to acquire the skills necessary to maximize the likelihood of success. To act is the job of an agent. The agent performs an action by first learning about the context in which it will occur. The agent keeps itself in a certain mental state to learn about its surroundings. The agent acquires knowledge of its surroundings via exploration and experimentation. The agent uses the reward function to gain knowledge of its surroundings. The agent receives either a positive reward or a negative punishment based on the acts it does. The agent's goal is to maximize positive reinforcement and minimize detrimental feedback. Human experts in the application domain are unnecessary for reinforcement learning. Reinforcement learning is useful in many contexts, including self-driving automobiles (Prasad et al. 2019).

2.1.3   Learning Methods

Multiple methods of education are discernable. You can classify the various methods of learning as:

1) Generalization Learning and
2) Discovery learning

We classify various methods of instruction as follows:

1) Rote learning
2) Learning by taking advice
3) Learning by example
4) Learning in problem-solving

The following categories have been established for more study:

1) Supervised Learning
2) Unsupervised Learning
3) Reinforcement Learning
4) Semi-Supervised Learning

To infer a function from data that have previously been characterized is the purpose of supervised learning. The training data consist of instructional examples. Instances are represented by pairs, each of which has a class and an input value. Supervised learning algorithms take a training set of examples and utilize them to infer a function that can be used to map further instances. In the best-case situation, the algorithm can reliably assign accurate labels to newly encountered instances. The challenge of unsupervised learning in ML is to classify data without labels into groups with a predefined degree of similarity (Bharti et al. 2021). The lack of a clear error signal while looking at provided examples prevents the learner from focusing on the best approach. Of the aforementioned learning problems, reinforcement learning is the broadest. Instead of being instructed about what to do by a superior, a reinforcement learning agent must learn by experience. To solve a problem, a learner employs reinforcement learning, in which he takes action on his surroundings and receives feedback in the form of a reward or punishment. The system discovers the best plan of action by making mistakes.

According to research, the most beneficial plan of action may consist of a series of actions carefully crafted to maximize returns (Bottou 2010). There is a lot of unlabeled data but not a lot of tagged data in many real-world learning domains like biology or text processing. As a rule, it takes a lot of effort and money to create data that have been appropriately categorized. As a result, SSL refers to the method of learning from a combination of labeled and unlabeled information. This kind of learning combines features of both supervised and unsupervised methods. Semi-supervised learning excels in scenarios when there is more unlabeled data available than labeled data. This occurs when the cost of collecting data points is low, while the cost of obtaining labels is high.

2.1.4 E-Learning with Machine Learning

E-learning refers to the process of acquiring knowledge via the use of electronic means of communication and dissemination. Further limiting its scope to the internet, Rosenberg defined e-learning as "the use of Internet technologies to deliver a broad array of solutions that enhances knowledge and performance." Helping students learn and grow as a result of their experiences is what is commonly referred to under many different names, including e-learning, virtual learning, web-based learning, computer-assisted learning, Internet learning, distributed learning, and distance teaching. Through the context of online learning, the student interacts with the course's teacher and other students in asynchronous, collaborative activities that take place via the Internet. Several researchers over the past few years have examined many different aspects of e-learning, with a primary focus on creating novel strategies for content presentation and enabling greater student–teacher interaction and collaboration (Alharbi and Doncker 2019).

The development of reskilling and upskilling and the supplementation of the conventional educational system are all attributable to the advent of e-learning. E-learning uses web technologies to make a structured, learner-focused, interactive, and facilitated learning environment available to anyone, anywhere, and at any time. An effective e-learning platform will have some sort of adaptive mechanism. The primary goal of adaptive systems is to facilitate individualized digital instruction. The foundation of meaningful learning is a constructivist method of conceptually modeling an individual's existing body of knowledge and experience with an eye toward adaptation. This latter method relies on ML algorithms and has made great strides toward mimicking a human expert's performance. When daily learning patterns are included in the model, desirable results are attained, giving e-learning platforms a distinct advantage in the market. Efficient incorporation of personalization into educational websites requires an e-learning system to dynamically detect the user model. If we use semantic analysis of e-content and global ontologies in an inferential manner, we can construct a learner model that stores our inferences about the user's perspective on the e-content (Anjaria and Guddeti 2014).

2.1.5 Need for Machine Learning

What if we used ML instead of programming our computers to do work? Programs that learn and develop based on their "experience" may be necessary due to the complexity of the issue and the need for adaptability:

1) **Tasks performed by animals/humans**: There are many things that we humans do regularly, but we do not reflect deeply enough on our processes to extract a clearly defined program. Tasks like driving, voice recognition, and visual comprehension are all examples. After being exposed to enough training instances, state-of-the-art ML systems and programs that "learn from their experience" perform well in all of these areas.

2) **Tasks beyond human capabilities:** The analysis of very large and complex data sets is yet another broad category that can benefit from ML techniques; examples include astronomical data, converting medical archives into medical knowledge, weather prediction, genomic data analysis, web search engines, and electronic commerce. As more and more information is captured digitally, it is becoming clear that vast troves of valuable knowledge lie dormant in databases that are incomprehensibly massive and complicated for humans to comprehend. The combination of learning programs with the almost infinite memory capacity and ever-increasing processing power of computers opens up new vistas in the field of learning to find significant patterns in big and complicated data sets.

2.2 Methodological Advancement of Machine Learning

Students come from a variety of backgrounds and have varying learning styles and pedagogical requirements. The primary goal of an adaptive e-learning system is to identify the specific requirements of each learner and then, following the training process, supply that learner with content that is tailored to his or her specific needs. Using ML and deep learning (DL) models with the right dataset may make the training process of an e-learning system more robust. In addition, an efficient intelligent mechanism is needed to automatically classify this content as belonging to the learner's category in a reasonable amount of time. This reduces the time spent by the learner searching through the vast amounts of content available within the e-learning environment to find something relevant to their specific needs. By doing so, we can tailor the information to each user. However, a multi-agent approach can be used in an e-learning system to tailor e-content to each student by tracking how they engage with the system and gathering data on their preferred methods of instruction (Araque et al. 2017).

2.2.1 Automatic Learner Profiling Agent

The method of developing a conceptual model through learner profiling allows the system to discover the learner profile and implicitly classify learners. E-content can then be provided to students based on their categories, and teachers can reach out to students who are less interactive to help them understand the material and pinpoint where their problems lie. The learner model is essential to adaptive e-learning because it describes the learner's characteristics and provides recommendations based on those descriptions. A learner's profile is automatically derived using their learning style, currently enrolled courses, and knowledge. Several AI approaches can help with this by grouping online course participants based on their shared characteristics and prior experiences.

To deduce the user profile from user activity, the intelligent profiling agent makes use of these data. Because the agent uses expert knowledge to assess user features, it might be considered an expert system (Chikersal et al. 2015). To select the most appropriate service to suggest, another agent is employed. The agent mimics the human expert's teaching style by using ML to figure out how to best serve the learner in question. Several factors, such as the probability or certainty factor in representing the user's behavior, are dependent on the model used to describe behavior, profile, and services. While the idea of context is utilized to model the course material, the representation of a student's knowledge state may be done using probabilistic logic. In this way, we have a formalized definition of the content model, the student model, and the instructional strategy. Each student's profile in the profiling system has a record of their academic and social background. The user's history has been modeled to create the model of the pupil. Students will receive customized course materials based on their unique learning plans, which are generated using data from both the student and content models.

2.2.2   Learning Materials' Content Indexing Agent

The cognitive level at which a learner acquires information about a subject dictates the learning materials that should be used to teach that subject. E-content resources may be selected for optimal performance in the learning process by considering the specific habits of each student. A model that may propose services in which each student is linked to a list of necessary ideas is needed. Learners can select useful resources, both physical and digital, by following the approach (Fancellu et al. 2016). This procedure necessitates specific conditions, including the retrieval of relevant materials via appropriate indexing. Web-based e-learning solutions that adhere to industry standards like SCORM are abundant, i.e. the reusable and interoperable Learning Content Object Reference Model (SCORM) specification. To ensure compatibility among SCORM conformant systems, SCORM offers a specific method for developing Learning Management Systems and instructional material.

2.2.3   Adaptive Learning

Learning in which proficiency is adapted about either one's surroundings or one's learning task is known as adaptive learning. Data, history, and experience are the building blocks of education. When it comes to various forms of education, a technique that works wonders in one context might not be the best option. Humans employ a wide variety of tactics for learning new material and tackling challenges. Learning a new language requires a different approach than studying for a math test. The learning issue, or what is to be learned and the goals of the learning process, is intrinsically linked to the learning procedure itself. Therefore, comprehension of the issue is necessary for the selection of a learning approach. The first step in adaptive learning is to identify the nature of the learning problem, and the second is to select, in real-time, the most appropriate strategy for addressing that problem. More than just trying out some novel approaches, enlisting the aid of a few eager students is needed. The best method can be decided upon after careful data selection. Adaptive learning is a strategy that modifies learning environments based on a set of policies. It is conditional on the specifics of the input circumstance and the user's real surroundings (Vanschoren et al. 2014).

Predicting the efficacy of ML algorithms, ranking learning algorithms, and selecting the most appropriate classification approach are all topics of study in the fields of ML and data mining. The algorithm selection problem is investigated in this chapter. Several algorithms exist that can approximate or precisely solve a problem. Find the best algorithm(s) and suggest them to the user.

In reality, the data characteristics of different datasets influence the performance of a classifier in different ways, which accounts for the seen variation in results. According to the extensive empirical results presented by (Romero et al. 2013), "decision trees do well on credit datasets, k-nearest neighbors excel on problems concerning images." There is widespread agreement that data characteristics should be considered when selecting an appropriate algorithm. This is extremely useful for practical purposes.

2.2.4 Proposed Research

The proposed classification algorithm recommendation method differs from the aforementioned work in how it characterizes a dataset and how it handles datasets that are similar to the one being characterized. In more detail, the k-nearest neighbor (k-NN) method finds the k datasets that are most similar to the new dataset (Jankowski 2013). The algorithms that perform best on datasets that are most similar to the one in question are then suggested as a place to begin addressing the data concealment issue. Experimental results on 38 UCI datasets with nine different classifiers corroborate the efficiency of the proposed method.

2.2.5 Multi-Perspective Learning

Information and knowledge about the system, gathered from a variety of sources, must be recorded in its entirety. If you want to maximize your learning, you need to take use of as much data as possible. There is a new viewpoint to be gained with each new piece of data. In this regard, some points of view are crucial while others are less so. When information regarding other views is lacking, making decisions can be tough. This means that there is a chance of learning gaps. To effectively make decisions that take into account several factors, it is vital to acquire knowledge from a variety of sources (Bhatt et al. 2012).

Learning from "knowledge and information obtained and created from many viewpoints" is what we mean when we talk about multi-perspective learning. To achieve multi-perspective learning, it is necessary to record data, system characteristics, and connections while considering several viewpoints. The multi-perspective approach to learning may increase the complexity of education in general, but it also offers many more educational opportunities. One's point of view may be defined as their current mental or factual condition. All the information from the current window pertains to a certain issue domain. Multi-perspective learning is a process wherein individuals bring such as P_1, P_2, ..., and P_n are combined to help in decision-making.

2.2.6 Machine Learning Recommender Agent for Customization

2.2.6.1 E-Learning

The model takes into account both the students' tastes and the content of the online courses. We develop a recommender agent using an ML model to establish a connection between a learner's preferences and an appropriate e-learning category and material (Prudencio et al. 2011). In this study, we employ Naive Bayes and K-Means, two ML models, to categorize the course contents into different types of e-learning resources. The system may operate in both an unsupervised learning mode and a supervised learning mode, the latter of which makes use of previously collected data for training purposes. The concept allows the system to provide recommendations for learning resources that are tailored to each student. The data were used to compare the efficiency of two ML models.

2.2.7 Data Creation

The model is implemented and assessed using a custom dataset compiled from user survey data collected via a Google form. Ninety-six resources representing various types of education are gathered with the help of college students. All student requests have been entered into a 12 by 1500 grid form. To keep track of how often certain concepts appear in the material the student has selected, the values in the columns are used, e.g. Material = {Web application, C-Programming, Python Programming, Java ...etc.}, Category = {Software, Artificial Intelligence, Multi-media, Networks .... etc.}. After collecting the data, in this step, the dataset undergoes preprocessing and cleaning. The information must be transformed from its current textual format into a numerical one (Alkawaz et al. 2018). The information has been used to test the ML models' performance. Various Python libraries are used for preliminary data cleaning and analysis, and the Scikit learn package is used to run experiments.

2.2.8 Naïve Bayes model

The NB model is implemented as a Bayesian supervised learning classifier. Classifying the e-learning resources for an individualized e-content recommendation determines how closely related the concepts are to the materials. Bayes' theorem was used to determine the posterior probability $(G|L)$ of class (Content; Category), as follows:

$$P\left(\frac{G}{L}\right) = \frac{P\left(\dfrac{L}{G}\right) * P(G)}{P(L)}$$

In this formula, the likelihood of a class is denoted by $P(L|G)$, and the prior probabilities of content (G) and predictor category $(P(L))$ are used to determine the likelihood of a class (L) given content (G). The conditional probability between the categories and the content is calculated based on the data's past.

The procedure uses a data set to generate a numeric value "LabelEncoder().fit_ transform()," which permits a string to be converted into a numerical mode that is machine-readable. The data will then be represented using the Naive Bayes model's preferred method.

For example:
The function "datafram['Term']=number.fit_transform(datafram['Term'])" has been used which represents the term "Software" with the number "10," "Multimedia" with the number "8" and so on.

2.2.9 K-Means Model

K-means, a form of clusterization that classifies learning objects into k categories using learners' past data, is an unsupervised classification algorithm with limited labeled training data. The objects are clustered so that their average distance from the cluster center is minimized. There are 12 words and 96 materials represented in the collected data. The challenge for each material is to determine which of the two centers it most closely resembles, assuming that all quantities, including those of time and matter, are real scalars. Clusters of points with axes are defined by their centroid. Reassigning materials to clusters takes into account their proximity to the centers of existing clusters, or, in another way, their similarities to the materials already in those clusters (Badrinarayanan et al. 2017). K random terms $\{t_1, t_2, ..., t_k\}$ are selected as seeds. After a predetermined number of cycles, the centroid positions will remain stable.

Concerning every substance (m_i): Assign mi to the nearest Term t_j (cluster) such that distance (m_i, t_j) is minimal. Next, update the seeds to the centroid of each cluster.

For each cluster t_j, $t_j = (m_j)$

Having 12 terms where each term represents a cluster and each cluster contains eight materials was previously illustrated. The data set must be transformed into a numerical format so that it can be used as input in our model. The function "random. Uniform ()" has been employed, allowing us to produce random numbers (here, eight numbers for each cluster, yielding 96 materials for all clusters) in a range from 0 to 1, to which a fixed number is subsequently appended.

For example:
The function "data ['Software']= np.random.uniform(0,1,8)+2" has been used, a visual representation of the k-means clustering algorithm's partitioning of the learning material adaptability model into clusters. When introducing a brand-new set of topics and a fresh set of student profiles, unexpected complications might occur. To account for these variations, the frequency table database is being dynamically updated. Less inaccuracy will be introduced into the system's predictions of relevant material based on a user's profile (Bayar and Stamm 2018).

2.3 Machine Learning on Time Series Analysis

Time series are mathematical constructs that denote a series of data ordered and indexed by time. In addition, a time series is a collection of measurements taken at regular intervals in the past, called y_t, each of which has a real value and a time stamp. The importance of the data's ordering across time is what sets time series apart from other types of information. Time series values are typically collected by keeping track of some process with time, with measurements taken at set intervals. One mathematical definition of a time series is

$$y_t = \left\{ y_1, y_2, ..., y_n \right\} \tag{2.1}$$

Since a time series can only be observed a finite number of times, the underlying process can be assumed to be a set of random variables in n dimensions. In addition, it is beneficial to assume the underlying process is a stochastic one, which allows for an infinite number of observations. When a mathematical function is applied to observe time series data, $y_t = f$ (time), then the series is said to be deterministic. Additionally, the time series is said to be nondeterministic or stochastic, when data are observed by the mathematical function $y_t = f$ (time, ϵ), where ϵ is the random term. Furthermore, stationarity is an important feature in time series. Properties (such as statistical properties) of a stationary time series remain constant across time (Birajdar and Mankar 2013).

Moreover, when we talk about the statistical properties, we are referring to things like the time series' mean value, auto-correlation, and variance. Univariate time series (UTS) and multivariate time series (MTS) are the two primary classifications of time series data. One way to think about an MTS is as an infinite series of numerous UTSs. Both UTS and MTS are widely available now because of the proliferation of numerous real-world applications and human activities that produce them. Everything from signatures and voice recordings to stock market data and medical signals is an example of biometrics. Time series data involve different sources such as monthly financial data, yearly birth rate data, hourly internet data, and annual temperature data.

Time series data come from a wide variety of modern-day sources. From biological signals and weather recordings to stock market rates and countless other sources, data are constantly being generated from a wide variety of human activities and practical applications. The time series data are typically gathered through the use of sensors that record physical quantities and then convert those readings into signals that can be easily understood by computers or humans. There has been a surge in interest, both in academia and in the field, in modeling non-temporal data as time series. It has been demonstrated that it is possible to transform data types as time series signals, including those used in video retrieval, picture retrieval, handwriting recognition, and text mining jobs (Dang et al. 2019).

There is a lot of interest in studying how to extract useful information from the massive amounts of time series data produced by a variety of applications (whether they are strictly temporal or simply sequential); nonetheless, it remains a difficult issue to solve. The primary objective of time series analysis is to derive actionable statistics and other features from time series data. Due to the inherent temporal ordering of time series, the study of these problems is more challenging than other data mining tasks. "Time series data mining" means "mining" data from time series for insights. Particularly, time series data mining is a product of math and computer science coming together, ML, AI, and statistics with time series data (Flach 2012). When analyzing time series data, scientists are on the lookout for things like anomalies, commonalities, and natural groupings. Classification, grouping, prediction, segmentation, anomaly detection, representation, and indexing are all typical time series data mining activities. This dissertation focuses on issues surrounding the representation, categorization, and prediction of time series, an issue with representing, categorizing, and forecasting time data.

2.3.1 Time Series Representation

Since learning directly from time series data is typically inefficient and laborious, the fundamental concern in time series mining is how to represent the time series data. Proposing algorithms that directly work on raw time series is computationally costly due to the large dimensionality of the time series. All real-world time series data are extremely highly dimensional, which is a major source of worry for the representation problem. Additionally, several time series data mining learning systems are predicated on time series representation and modeling (Jaiswal and Srivastava 2020). When dealing with time series data, the large dimensionality causes a dramatic drop in performance for non-trivial indexing and data mining techniques. Thus, transforming time series into reduced dimensionality form is a typical method for representing time series. Mathematically, the concept of time series representation can be defined as an operation T which transforms raw time series Y_t into another time series form $Y\hat{}t$ or a scalar $xT(Yt)$ such that $Y\hat{}t$ or $xT(Yt)$ summarizes a given pattern in Yt.

$$T\left(Y_t\right) \begin{cases} Y_t \\ \text{or} \\ x|\left(Y_t\right) \end{cases} \tag{2.2}$$

It may seem counterintuitive to substitute an approximation for the actual value of a measurement. However, in time series, the trends, forms, patterns, and themes are more important than the individual data points themselves. A proper high-level representation may define and capture these patterns, in addition to actively suppressing background noise. The primary goal of representation is to clearly and simply communicate the key aspects of the time series data. This also facilitates faster processing, space-saving storage, and noise cancellation. Numerous time series representation techniques have been developed to facilitate similarity search and data mining operations. Different types of these representations are time domain representations, frequency domain representations, and model-based techniques (Lloyd et al. 2013).

The temporal or sequential structure of time series data is maintained via representation techniques that operate in the time domain. This is done by calculating a global or local approximation of a feature. In addition, data-adaptive and non-data-adaptive temporal domain approaches exist. To reduce the amount of error introduced when reconstructing the entire dataset, data-adaptive techniques will select a single representation to use for all time-series instances. In contrast, the non-data-adaptive approaches use local attributes to build an approximation of the true representation. Also, transformations like the Fourier transform and the wavelet transform are used to turn the time series into features in the frequency domain in the frequency domain representations. On top of that, model-based representations presumed that a model was responsible for creating the observed time series data. Regression models and other approaches are used by these representation techniques to uncover the corresponding parametric form of the underlying model.

There have been several proposed methods for decreasing the dimension of time series data. The majority of these techniques involve shortening a time series from l elements to n elements. The fastest and most straightforward method for reducing the number of dimensions is sampling. In addition, we present piecewise approximation methods, which accomplish the same goal by approximating the time series using subsamples and an aggregate function. There are several different piecewise methods described in the literature, but the most common ones are the piecewise aggregate approximation (PAA), the piecewise linear approximation (PLA), the adaptive piecewise constant approximation (APCA), and the perceptually relevant spots. By varying the length of the segments over time, the APCA is essentially a dynamic variant of the PAA technique.

Moreover, each time series segment is represented by the segmented sum of variation (SSV), and dimensionality reduction is achieved via bit-level approximation. In addition, the PIP approach, which has only recently been developed, may be used to retain the most important aspects of financial time series data. Given the transient nature of financial data, PIP succeeds where other piecewise reduction methods fail by collecting key information. In addition, signal processing techniques that transform signals from the time domain to the frequency domain can have an impact on particular representations of time series data. Transforms like the Discrete Fourier Transform (DFT) and Discrete Wavelet Transform (DWT) are typical additionally, a comprehensive investigation into the various numerical time series formats.

2.3.2 Time Series Classification

One type of supervised learning job that is commonly used in ML and data mining is classification. Objects are classified by a process of labeling them with predetermined categories. Specifically, classification is the method of using previously labeled data as training to make predictions about new data that have not yet been analyzed. A classifier is a method for implementing classification, wherein a mathematical function is used to map fresh input data to a category. Classification error is commonly used as a measure of a classifier's efficacy. In most cases, three stages are required for categorization. The first stage involves employing a training technique to construct a classifier from scratch using a fixed labeled data set. Each time series instance requires a known class label to be used in the training process. The categorization criteria that assign a unique identifier to each time series object have since been established. The second stage involves validating the trained classifier with a test dataset. Each time series example in the test set has to have its corresponding class label recognized. By comparing the predicted labels for the test set to the actual test class labels, the performance of the classifier may be verified. If the classifier's performance falls short of expectations, it will need to be retrained. Parameter tweaking and adjusting the learning rate are two more ways to improve the classifier's efficiency. Finally, the unlabeled time series instances may be identified using the resulting classifier, which has shown to have excellent validation accuracy on the test set.

To solve the problem of labeling data points in a time series with one of several possible classes, it is common to practice employing a supervised classifier, which can be represented as a training set of univariate time series sequences. UT1, UT2, ..., UTn have been assigned to c number of class labels cl1, cl2, ..., clc where each UTi = (UTi1, UTi2, ..., UTil) of length l. There have been a plethora of methods developed over the past few decades for tagging time series. Time series classification techniques are typically sorted into one of three categories based on the approach used to categorize the data: distance-based, feature-based, and model-based. K-nearest neighbors (k-NN) calculated with Euclidean distance and time-varying warping are used by distance-based techniques for data classification. The first standards were established using these methods, with k-NN with DTW emerging as the undisputed leader over the past 10 years. By contrast, feature-based methods first transform the time series before extracting features that are more representative and distinguishable from the time series and then classify the data with a 1-NN classifier. The bag of features framework (TSBF), the bag of symbolic Fourier approximation symbols (BOSS), and the bag of symbolic Fourier approximation symbols (BOSS) are all recent examples of feature extraction frameworks. Moreover, model-based methods expect some underlying model to correctly characterize a specific class. When it comes to classifying time series, ensemble methods are typically used, and the most successful of these is the flat collective transform-based ensemble (COTE). Extensive pre-processing and feature extraction from time series were required for classification using any of the aforementioned methods. The data must be highly dynamic, noisy, multi-dimensional, and long-lasting, and both time series must be of the same length for distance-based methods to be effective. However, DTW has not been widely adopted for time series data mining projects due to its computational overhead and inability to meet the triangle inequality, despite its long history of use. There is a strong relationship between the quality of a classifier's performance and the care with which its features were developed. Remember that time series gathered from practical applications are typically not able or very difficult to be represented by a generative model, which greatly restricts the utility of model-based approaches.

2.3.3 Time Series Forecasting

Accurately predicting future occurrences, like the amount of rain that will fall next year or the volatility of the stock market, is a difficult topic in both applied and theoretical time series research. Time series analysis has several possible uses, including but not limited to the following: statistics, econometrics, signal processing, pattern identification, mathematical finance, climate prediction, electrocardiogram (ECG), control engineering, astronomy, etc. Time series data can be difficult to analyze and model because of its vastness, complexity, non-stationarity, volatility, and inability to be predicted in advance. Time series analysis approaches have been created to help academics and statisticians deal with these problems.

1) The goal of time series analysis is to determine the probability rule that best fits the data. A parametric model is one in which all the variables, including the probability rules, are known, except for a small number of finite-dimensional parameters. Parameters in infinite-dimensional space also designate a model as a nonparametric model. Furthermore, numbers referred to as parameters, primarily the mean and standard deviation, describe theoretical distributions of time series data. Parametric models are convenient for time series analysis since they are straightforward and demand little in the way of computational resources. Parametric time series models have a deep theoretical foundation and several practical uses, including supplementary research in the disciplines of economics and ecology. Estimating the parameters of the assumed probability distribution allows us to create a model for time series data using a parametric approach.

2) In addition, parametric and nonparametric approaches exist within the realm of time series analysis. Parametric methods presume that the fundamental stochastic stationary process has a well-defined structure that can be characterized with a manageable set of parameters (as in the case of the moving average (MA) or autoregressive (AR) model). Parametric methods primarily focus on determining the best values for the model parameters of a time series that characterizes a random process. Instead of making assumptions about the stationary process's structure, nonparametric approaches make an explicit evaluation of the process's spectrum or covariance (Luo et al. 2019).

3) In addition, the recent decade has seen a rise in the use of ML techniques for time series forecasting. The key to its success is its capacity to learn from mistakes and become better as it goes through iterations. ML methods, a broad category that includes models like neural networks (NN), support vector regression (SVR), regression trees, the Gaussian process (GP), and long short-term memory (LSTM), have all been used for time series forecasting. In particular, NNs are used to increase the reliability of the forecasts made. Temporal series forecasting has been used with neural networks ever since they were first developed, and many designs from the simple multilayer perceptron (MLP) to the more complicated time delay network (TDN) and recurrent neural network (RNN) have all been put to use (RNN).

2.4 Conclusion

Extensive multi-agent-based ML is used for online education. The suggested system is made up of three components: an autonomous profiling agent, a content indexing agent, and an ML recommender model. The students are sorted into appropriate e-learning categories based on their profiles and patterns of online activity. Learning resources are categorized using ML models according to the student's preferred learning approach and personal preferences. The Naive Bayes and K-Means algorithms and how to implement them are described in depth. The findings indicate that the K-Means model is more precise than the Naive Bayes one. Meta-goal learning is to advise the user on the learning algorithm to employ for every given data-mining project. When designing a meta-learning system, it is crucial to accurately define the features of the dataset to predict the efficacy of the learning algorithm. By identifying the best possible student, this framework boosts the reliability of the learning process. Experiments on meta-features reveal that the basic meta-features, including the number of attributes, the number of instances, the number of class labels, the maximum probability of a class, and the class entropy, are effective for classifier selection throughout the thirty-eight datasets tested.

Data in the form of time series are sequences of actual values that may be arranged in a meaningful way. The vast majority of data collected in any given industry may be represented as time series measurements because of the inclusion of a time stamp. Time series data have vast real-world uses, from measuring astronomical light intensities and physiological sensors to keeping track of commercial and financial transactions. Time series data analysis has resulted in a new breed of data mining problems (classification, clustering, prediction, representation, and indexing). In addition, researchers have spent years learning about time series and their peculiar characteristics. Classifications can be made more difficult depending on whether the data are continuous or discrete, univariate or multivariate, or if there are temporal dependence difficulties to consider. When faced with a novel time series classification challenge, the best possible solution is often found in developing a system that is specific to that issue. However, adapting this particular method to new issues may be extremely difficult and resource-intensive. In this chapter, we set out to address this problem by proposing a generic strategy that outperforms existing approaches and can be used to categorize time series datasets from any area. We are also inspired to use the suggested classification technique to categorize multivariate time series signals in addition to univariate ones.

Time series data are notoriously difficult to analyze and model because of their high dimensionality, complexity, and special features. The representation of time-series data is of great relevance for dimensionality reduction and information extraction. Further, time series patterns frequently display temporal displacement, size heterogeneity, the presence of arbitrarily repeating patterns, and the presence of local distortions/noise. Sometimes it is the smaller regional variances between groups that get the blame instead of the overall pattern. Because of these factors, values associated with a given timestamp have distinct meanings when applied to other time series. We refer to these events as intra-class variations.

In this dissertation, we also try to lower the dimensionality and number of time series data points.

Therefore, we offer a new representation approach that may be employed with existing classifiers. Additionally, we utilized our new representation approach for time series data to manage the intra-class variances in classification algorithms. Furthermore, we are driven to find a solution to the age-old challenge of accurately predicting future values. Given the proven efficacy of deep learning approaches in the field of image and signal processing for tasks such as face identification, object recognition, and masking, we focus here on forecasting many steps into the future using a single historical data set. Additionally, other non-computer vision fields have been examined by applying deep learning architectures.

Acknowledgment

This research was supported by "Building a Data/AI-based problem-solving system" of Korea Institute of Science and Technology Information (KISTI), and University of Science and Technology (UST), Daejeon, South Korea and my Professor, Min-ho Lee owe thanks for the support and excellent research facilities.

Conflict of Interest

The author declares that they have no known competing financial interests or personal relationships that could have appeared to influence the work reported in this chapter.

References

Alharbi, A.S.M. and de Doncker, E. (2019). Twitter sentiment analysis with a deep neural network: an enhanced approach using user behavioral information. *Cognitive Systems Research* 54: 50–61.

Alkawaz, M.H., Sulong, G., Saba, T., and Rehman, A. (2018). Detection of copy-move image forgery based on discrete cosine transform. *Neural Computing and Applications* 30 (1): 183–192.

Anjaria, M. and Guddeti, R.M.R. (2014). A novel sentiment analysis of social networks using supervised learning. *Social Network Analysis and Mining* 4 (1): 1–15.

Araque, O., Corcuera-Platas, I., Sánchez-Rada, J.F., and Iglesias, C.A. (2017). Enhancing deep learning sentiment analysis with ensemble techniques in social applications. *Expert Systems with Applications* 77: 236–246.

Badrinarayanan, V., Kendall, A., and Cipolla, R. (2017). Segnet: a deep convolutional encoder-decoder architecture for image segmentation. *IEEE Transactions on Pattern Analysis and Machine Intelligence* 39 (12): 2481–2495.

Bayar, B. and Stamm, M.C. (2018). Constrained convolutional neural networks: a new approach towards general purpose image manipulation detection. *IEEE Transactions on Information Forensics and Security* 13 (11): 2691–2706.

Bharti, R., Khamparia, A., Shabaz, M. et al. (2021). Prediction of heart disease using a combination of machine learning and deep learning. *Computational Intelligence and Neuroscience* 11: 1–3.

Bhatt, N., Thakkar, A., and Ganatra, A. (2012). A survey and current research challenges in meta-learning approaches based on dataset characteristics. *International Journal of Soft computing and Engineering* 2 (10): 234–247.

Birajdar, G.K. and Mankar, V.H. (2013). Digital image forgery detection using passive techniques: a survey. *Digital Investigation* 10 (3): 226–245.

Bottou, L. (2010). Large-scale machine learning with stochastic gradient descent. *Proceedings of COMPSTAT'2010* (August 22–27, 2010), pp. 177–186. Physica-Verlag HD: Paris.

Chikersal, P., Poria, S., Cambria, E. et al. (2015). Modeling public sentiment in twitter: using linguistic patterns to enhance supervised learning. *International Conference on Intelligent Text Processing and Computational Linguistics*, Cham (April 14–20, 2015), pp. 49–65. Springer.

Dang, L.M., Hassan, S.I., Im, S., and Moon, H. (2019). Face image manipulation detection based on a convolutional neural network. *Expert Systems with Applications* 129: 156–168.

Dang, L.M., Min, K., Lee, S. et al. (2020). Tampered and computer-generated face image identification based on deep learning. *Applied Sciences* 10 (2): 505.

Fancellu, F., Lopez, A., and Webber, B. (2016). August. Neural networks for negation scope detection. *Proceedings of the 54th Annual Meeting of the Association for Computational Linguistics* (August 7–12, 2016), pp. 495–504. Volume 1: long papers.

Flach, P. (2012). *Machine Learning: The Art and Science of Algorithms that Make Sense of Data*. Cambridge University Press.

Jaiswal, A.K. and Srivastava, R. (2020). A technique for image splicing detection using hybrid feature set. *Multimedia Tools and Applications* 79 (17): 11837–11860.

Jankowski, N. (2013). Meta-learning and new ways in model construction for classification problems. *Journal of Network & Information Security* 4 (4): 275–284.

Kumar, N., Narayan Das, N., Gupta, D. et al. (2021). Efficient automated disease diagnosis using machine learning models. *Journal of Healthcare Engineering* https://doi.org/10.1155/2021/9983652.

Lloyd, S., Mohseni, M., and Rebentrost, P. (2013). Quantum algorithms for supervised and unsupervised machine learning. *ArXiv Preprint ArXiv* 1307: 0411.

Luo, S., Peng, A., Zeng, H. et al. (2019). Deep residual learning using data augmentation for median filtering forensics of digital images. *IEEE Access* 7: 80614–80621.

Malakar, A.K., Choudhury, D., Halder, B. et al. (2019). A review on coronary artery disease, its risk factors, and therapeutics. *Journal of Cellular Physiology* 234 (10): 16812–16823.

Masetic, Z. and Subasi, A. (2016). Congestive heart failure detection using random forest classifier. *Computer Methods and Programs in Biomedicine* 130: 54–64.

Patel, S. and Patel, A. (2016). Big data revolution in the health care sector: opportunities, challenges & technological advancements. *International Journal of Information* 6 (1/2): 155–162.

Prasad, R., Anjali, P., Adil, S., and Deepa, N. (2019). Heart disease prediction using logistic regression algorithm using machine learning. *International Journal of Engineering and Advanced Technology* 8 (3S): 659–662.

Prudêncio, R.B., De Souto, M.C., and Ludermir, T.B. (2011). Selecting machine learning algorithms using the ranking meta-learning approach. In: Norbert Jankowski, Włodzisław Duch & Krzysztof Grąbczewski, (Eds.), *Meta-Learning in Computational Intelligence*, 225–243. Berlin, Heidelberg: Springer.

Qiu, J., Wu, Q., Ding, G. et al. (2016). A survey of machine learning for big data processing. *EURASIP Journal on Advances in Signal Processing* 2016 (1): 1–16.

Romero, C., Olmo, J.L. and Ventura, S. (2013). A meta-learning approach for recommending a subset of white-box classification algorithms for moodle datasets. *Educational Data Mining 2013* (July 6–9, 2013). Memphis, Tennessee, USA.

Sarker, I.H. (2021). Machine learning: algorithms, real-world applications & research directions. *SN Computer Science* 2 (3): 1–21.

Vanschoren, J., Brazdil, P., Soares, C., and Kotthoff, L. (2014). Meta-learning & algorithm selection workshop at ECAI 2014.

3

Time-Series Forecasting for Stock Market Using Convolutional Neural Network

Partha Pratim Deb[1], Diptendu Bhattacharya[1], Indranath Chatterjee[2], and Sheetal Zalte[3]

[1] Department of Computer Science and Engineering, National Institute of Technology Agartala, Agartala, Tripura, India
[2] Department of Computer Engineering, Tongmyong University, Busan, South Korea
[3] Department of Computer Science, Shivaji University, Kolhapur, Maharashtra, India

3.1 Introduction

Time series prediction has drawn a lot of attention from many different research fields as a key area of dynamic data analysis. Many forecasting techniques, including local and global models and univariate and multivariate models, were created in this discipline. While multivariate approaches often analyze the relationship between the time series, univariate forecasting methods simply read a time series' previous values to determine where it will go in the future. In order to anticipate the demand for a large number of similar time series, global approaches are useful. In these methods, model parameters are collectively evaluated based on the entire set of time series, while in local methods, parameters are, in my view, evaluated for each individual time series. Ye and Dai (2021) have applied the convolutional neural network (CNN), which benefits from its strength in extracting nearby features through multiple convolutional filters and studying illustration by fully linked layers and has recently been found to be extremely significant in application areas, such as financial services, energy, and rechargeable load.

Because of the importance of time series forecasting and the uses of CNNs in modeling, there is a growing interest in building CNNs for time series forecasting. Regarding model selection, the typical paradigm for creating a CNN is to first choose

Machine Learning Applications: From Computer Vision to Robotics, First Edition.
Edited by Indranath Chatterjee and Sheetal Zalte.
© 2024 The Institute of Electrical and Electronics Engineers, Inc.
Published 2024 by John Wiley & Sons, Inc.

and fix the hyper-parameters (i.e., neural architecture, learning rate, and training epochs), and then the community is trained using a variety of gradient descent optimization techniques. This helps make it rigid and extremely difficult to comply the hyper-parameter to create CNNs. So, this study's focus is on creating a strong CNN for accurately predicting time series data provided by Zhang et al. (2021).

This study suggests a method for predicting the stock's final price the next day, which is based entirely on the CNN in order to make stock price predictions that are more accurate. Lu et al. (2021) applied the CNN which can extract features from the stock data that it receives as the input. The Financial Time Series is a collection of regular observations of Financial Variable(s). Financial time series include, for instance, daily change rates, daily stock market index values, and daily commodity fees. The time series of financial events is generally disordered and noisy. A complex time is not necessarily linear and sensitive to preceding circumstances. Financial time series have varying statistical characteristics over time and are also noisy. The forecast cannot be made because of this reason. It is continually difficult to create the ideal prediction model that might capture the nonlinearity present in the time series. Hence, it is shown that forecasting financial time series is a challenging and difficult process. In comparison to my own stand-alone forecasting models, some investigators have demonstrated that an ensemble or composite forecasting model for time series prediction can perform better. A CNN is a particular instance of a neural network that uses one or more convolutional layers, typically together with a subsampling layer, which may also use one or more absolutely linked layers, as in a preferred neural network. CNNs are a type of neural network that has been enhanced for two-dimensional image facts. Nonetheless, they may be used for one-dimensional facts, text-based sequences, and time series data which is found from Durairaj and Mohal (2022).

Figure 3.1 indicates the cross-field evaluation among the financial technology and machine learning model. Machine learning algorithms for the forecasting of monetary market prices are analyzed in various research studies, despite the risk of over-fitting and its complexity. Investors now use sophisticated trading frameworks to help them make decisions about what investments to purchase and sell. Using an algorithm for the accurate forecasting of stock market prices is even more crucial for financial firms. It is difficult to understand without an issue that is similar to a black-field system when using algorithms based on machine learning to anticipate stock market charges provided by Anuradha (2021). CNNs have been used to forecast stock values in an increasing amount of research in recent years.

The reason for this method is as follows:

- Our goal is to determine whether or not a deep learning model can properly forecast future stock rate trends when using the gold rate, the gold volatility index, the crude oil rate, and the crude oil rate volatility index as inputs.

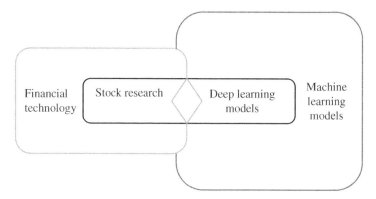

Figure 3.1 Stock market analysis based on existing research.

- We decide whether to buy, sell, or hold a specific stock using the extracted feature data and the trained version.

We divide the paper into four sections where in the methodology section, we discuss the method we use in this paper. Then, we discuss the method and formula we used in this paper. In the Materials section, we discuss the materials we used in the paper. In the Result and Discussion section, we have shown the RMSE value and the graph we get after calculating.

3.2 Materials

We have considered three unique sorts of indices for our experimental purpose, where three facts are set, KOSPI, TAIEX, and BSE-SENSEX. From the years 2015 to 2020, 365 days of records for each year are used for the inventory index. Data from January 1 to October 31 are used for training and those from November 1 to December 31 are used for testing for evaluating the effectiveness of the CNN.

3.3 Methodology

It is obvious that models exist that might serve as examples for both Deep Learning and Time Series Analysis in Machine Learning since they have evolved through various procedures. The method that is being suggested for this project relies around CNN models, which are excellent for forecasting.

3.3.1 The Convolutional Neural Network

CNN models are extremely detailed and intricate in design, and they are made up of the following layers:

- **Pooling (POOL)**: This technique causes a steady reduction in the spatial length of the representation in order to decrease the number of parameters and computations required inside the network. Each feature map is individually handled by the pooling layer.
- **Completely connected (FC)**: The fully connected layer is the last layer that functions on flattened inputs, where each input is coupled to every neuron. Neurons often exhibit sigmoid activation or softmax characteristics.
- **Convolutional layer (CONV)**: When it scans entry I with regard to its dimensions, the convolutional layer instills filters that perform convolutional operations. The clear out length F and stride S are included in the hyper-parameters. The output that is produced is known as the Features Map or Activation Layer.
- **Stride(S)**: The clear out hyper-parameter for strides. It indicates how many pixels the window will skip across following each action.

Convolution is particularly mathematically represented by the asterisk * symbol. The expression may be as follows: If input A and filter c are given.

$$B = A * c \tag{3.1}$$

1D- Convolution is given as,

$$g_i = c\left(\left(U * a\right)_i\right) \tag{3.2}$$

Figure 3.2 shows the architecture of the CNN model. A subsampling layer that reduces noise in discovered functions, such as the feature maps, may also be included in the CNN layer. The regression layer is used in conjunction with this layer.

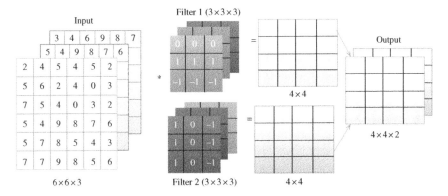

Figure 3.2 Architecture of the convolutional neural network.

3.4 Accuracy Measurement

This paper assesses the execution of the proposed strategy utilizing RMSE, which is characterized as taking after:-

$$\text{RMSE} = \sqrt{\frac{\sum\limits_{i=1}^{n}\left(\text{Actualvalue}_i - \text{Forecastedvalue}_i\right)^2}{n}}$$

where n signifies the number of days required to be forecasted.

3.5 Result and Discussion

From the above-proposed method, we get the below graphs and the RMSE values and graphs for BSE-Sensex, Taiex, and Kospi. In this section, we notice the proposed strategy to calculate the Sensex, Taiex, and Kospi insights from 2015 to 2020.

Tables 3.1 and 3.2 describe the CNN parameters in terms of model, epochs, number of layers, and activation function and show the RMSE.

Here, n indicates the variety of days that must be estimated. In Table 3.3, we observe the RMSE as an incentive for BSE Sensex, Taiex, and Kospi:

Table 3.1 Forecasted error rate of stock indices over 6 years.

Model	Epochs	Number of layers	Stock index	RMSE
CNN	100	5	BSE-SENSEX	43.718
CNN	30	5	BSE-SENSEX	58.391
CNN	100	4	BSE-SENSEX	54.219
CNN	30	4	BSE-SENSEX	48.301
CNN	100	5	TAIEX	102.854
CNN	30	5	TAIEX	203.518
CNN	100	4	TAIEX	145.612
CNN	30	4	TAIEX	187.815
CNN	100	5	KOSPI	12.728
CNN	30	5	KOSPI	20.451
CNN	100	4	KOSPI	16.601
CNN	30	4	KOSPI	18.312

Table 3.2 Prediction error rate over different activation functions.

Activation function	Stock index	RMSE
Sigmoid	BSE-SENSEX	48.619
Hyperbolic tangent	BSE-SENSEX	52.41
Relu	BSE-SENSEX	43.718
Leaky Relu	BSE-SENSEX	45.013
Sigmoid	TAIEX	122.723
Hyperbolic tangent	TAIEX	118.615
Relu	TAIEX	102.854
Leaky Relu	TAIEX	105.402
Sigmoid	KOSPI	16.315
Hyperbolic tangent	KOSPI	15.471
Relu	KOSPI	12.728
Leaky Relu	KOSPI	14.327

Table 3.3 RMSE values and average values of different methods.

Method \ Year	2015	2016	2017	2018	2019	2020
BSE Sensex	10.412	21.329	54.671	24.192	75.813	76.301
Taiex	46.823	87.902	119.631	45.734	198.114	118.921
Kospi	13.723	10.086	9.595	14.941	13.412	14.616

Then, in Figure 3.3, we display the actual and forecasted value for BSE Sensex from January to December 2015, where the data are found to be in an increasing and decreasing order every week of the year.

Then in Figure 3.4, we display the actual and forecasted value for BSE Sensex from January to December 2016, where the data are found to be in an increasing and decreasing order every week of the year.

Then in Figure 3.5, we display the actual and forecasted value for BSE Sensex from January to December 2017, where the data are found to be in an increasing and decreasing order every week of the year.

Then in Figure 3.6, we display the actual and forecasted value for BSE Sensex from January to December 2018, where the data are found to be in an increasing and decreasing order every week of the year.

Figure 3.3 Actual and forecasted value of 2015.

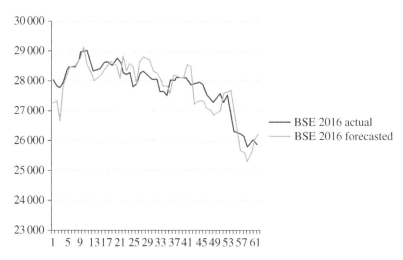

Figure 3.4 Actual and forecasted value of 2016.

Then in Figure 3.7, we display the actual and forecasted value for BSE Sensex from January to December 2019, where the data are found to be in an increasing and decreasing order every week of the year.

Then in Figure 3.8, we display the actual and forecasted value for BSE Sensex from January to December 2020, where the data are found to be in an increasing and decreasing order every week of the year.

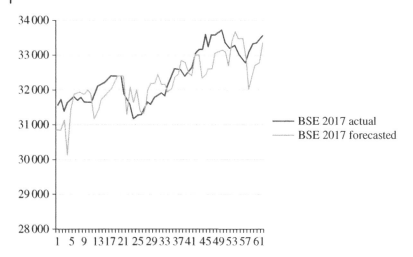

Figure 3.5 Actual and forecasted value of BSE 2017.

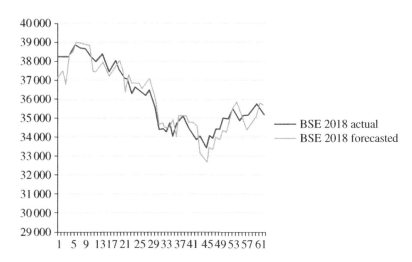

Figure 3.6 Actual and forecasted value of BSE 2018.

Table 3.4 indicates the comparison of RMSE values of the various years along with the average RMSE value in comparison with the earlier research (Chen and Huang, 2021; Chen et al. 2021; Dwivedi et al. 2021; Mukherjee et al. 2021; Nti et al. 2021; Wang et al. 2021) is the earlier research being followed with method BSE Sensex from the year 2015 to 2020.

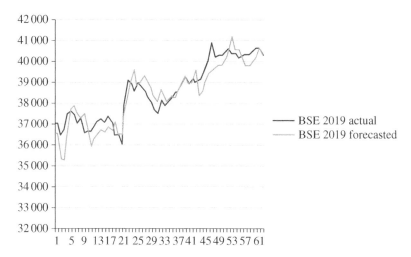

Figure 3.7 Actual and forecasted value of BSE 2019.

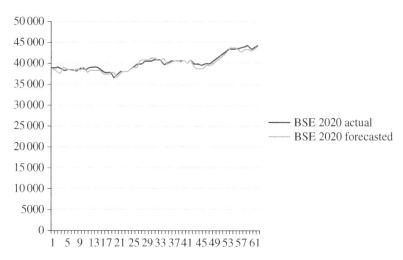

Figure 3.8 Actual and forecasted value of BSE 2020.

Then in Figure 3.9, we display the actual and forecasted value for TAIEX from January to December 2015, where the data are found to be in an increasing and decreasing order every week of the year.

Then in Figure 3.10, we display the Actual and forecasted value for TAIEX from January to December 2016, where the data are found to be in an increasing and decreasing order every week of the year.

Table 3.4 Comparison of the RMSE and the Average RMSE of BSE Sensex for Different Methods

Method \ Year	2015	2016	2017	2018	2019	2020
Chen and Huang (2021)	11.231	21.421	56.395	24.621	76.214	78.214
Chen et al. (2021)	11.321	21.223	55.365	22.567	77.214	75.214
Dwivedi et al. (2021)	11.321	19.214	55.214	25.987	76.218	78.362
Mukherjee et al. (2021)	10.231	20.148	54.214	23.521	76.214	77.654
Wang et al. (2021)	9.521	21.231	54.215	22.114	75.321	76.214
Nti et al. (2021)	11.245	22.658	56.972	23.217	77.621	78.261
Proposed model	10.412	21.329	54.671	24.192	75.813	76.301

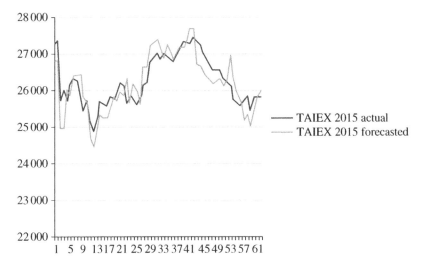

Figure 3.9 Actual and forecasted value of TAIEX 2015.

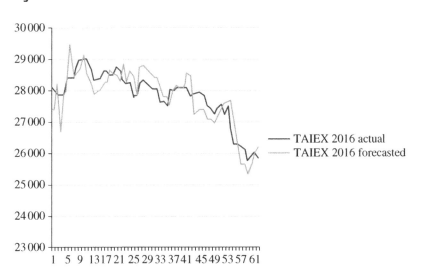

Figure 3.10 Actual and forecasted value of TAIEX 2016.

Then in Figure 3.11, we have display the actual and forecasted value for TAIEX from January to December 2017, where the data are found to be in an increasing and decreasing order every week of the year.

Then in Figure 3.12, we display the actual and forecasted value for TAIEX from January to December 2018, where the data are found to be in an increasing and decreasing order every week of the year.

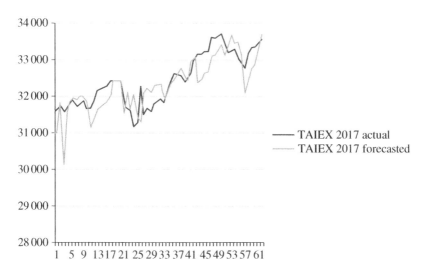

Figure 3.11   Actual and forecasted value of TAIEX 2017.

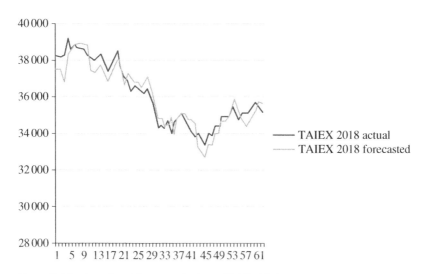

Figure 3.12   Actual and forecasted value of TAIEX 2018.

Then in Figure 3.13, we have display the actual and forecasted value for TAIEX from January to December 2019, where the data are found to be in an increasing and decreasing order every week of the year.

Then in Figure 3.14, we have display the actual and forecasted value for TAIEX from January to December 2020, where the data are found to be in an increasing and decreasing order every week of the year.

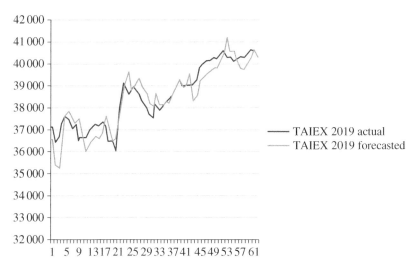

Figure 3.13 Actual and forecasted value of TAIEX 2019.

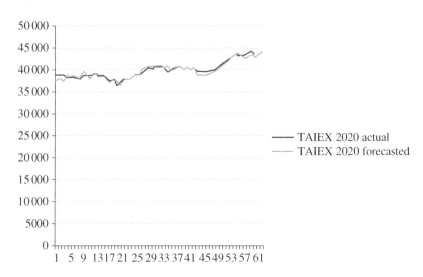

Figure 3.14 Actual and forecasted value of TAIEX 2020.

Table 3.5 indicates the comparison of RMSE values of various years along with the Average RMSE value in comparison with the earlier research studies (Chen's Fuzzy Time Series 1996; Huarng et al. 2007) is the earlier research studies being followed with the method Taiex from the year 2015 to 2020.

Then in Figure 3.15, we have discussed the actual and forecasted value for KOSPI from January to December 2015, where the data are found to be in an increasing and decreasing order every week of the year.

Table 3.5 Comparison of the RMSE and the average RMSE of Taiex for different methods.

Method \ Year	2015	2016	2017	2018	2019	2020
Chen and Huang (2021)	51.332	83.364	119.314	48.145	197.321	115.321
Chen et al. (2021)	50.332	83.222	119.321	45.369	196.321	113.124
Dwivedi et al. (2021)	48.367	84.166	117.321	46.321	197.32	113.211
Mukherjee et al. (2021)	49.332	84.329	120.325	45.397	197.326	114.985
Wang et al. (2021)	50.214	86.327	119.214	45.369	196.347	112.697
Nti et al. (2021)	48.321	82.321	118.321	46.311	196.321	111.212
Proposed model	46.823	87.902	119.631	45.734	198.114	118.921

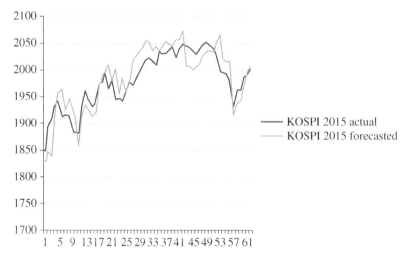

Figure 3.15 Actual and forecasted value of KOSPI 2015.

Then in Figure 3.16, we have discussed the actual and forecasted value for KOSPI from January to December 2016, where the data are found to be in an increasing and decreasing order every week of the year.

Then in Figure 3.17, we have discussed the actual and forecasted value for KOSPI from January to December 2017, where the data are found to be in an increasing and decreasing order every week of the year.

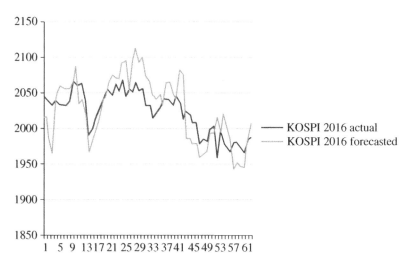

Figure 3.16   Actual and forecasted value of KOSPI 2016.

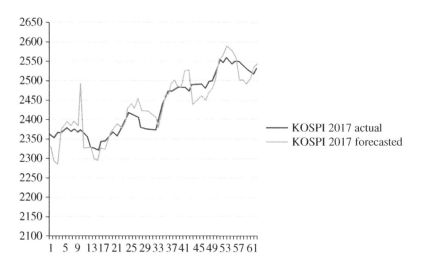

Figure 3.17   Actual and forecasted value of KOSPI 2017.

Then in Figure 3.18, we have discussed the actual and forecasted value for KOSPI from January to December 2018, where the data are found to be in an increasing and decreasing order every week of the year.

Then in Figure 3.19, we have discussed the Actual and forecasted value for KOSPI from January to December 2019, where the data are found to be in an increasing and decreasing order every week of the year.

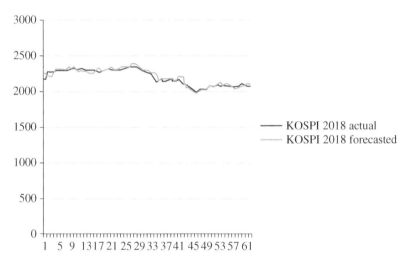

Figure 3.18 Actual and forecasted value of KOSPI 2018.

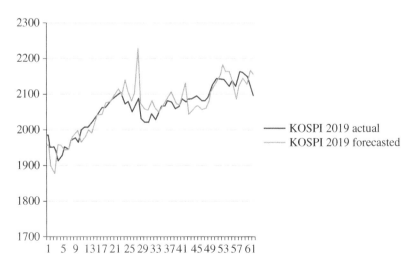

Figure 3.19 Actual and forecasted value of KOSPI 2019.

Then in Figure 3.20, we have discussed the actual and forecasted value for KOSPI from January to December 2020, where the data are found to be in an increasing and decreasing order every week of the year.

Table 3.6 indicates comparison of RMSE values of the various year along with the Average RMSE value in comparison with the earlier research (Chen's Fuzzy Time Series1996; Huarng et al. 2007) is the earlier research studies being followed with the method Kospi from the year 2015 to 2020.

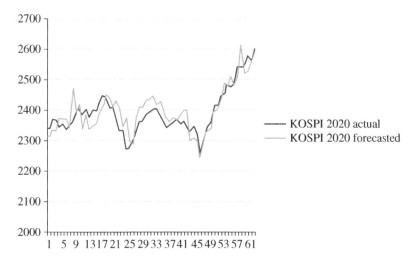

Figure 3.20   Actual and forecasted value of KOSPI 2020.

Table 3.6   Comparison of the RMSE and the average RMSE of Kospi for different methods.

Method \ Year	2015	2016	2017	2018	2019	2020
Chen and Huang (2021)	13.661	11.213	9.213	15.321	13.214	15.214
Chen et al. (2021)	12.321	2.315	9.213	14.216	15.321	14.115
Dwivedi et al. (2021)	13.214	9.321	8.001	16.321	12.369	14.215
Mukherjee et al. (2021)	13.623	9.321	8.321	12.621	13.214	14.213
Wang et al. (2021)	13.214	10.132	8.621	15.311	13.621	14.741
Nti et al. (2021)	14.211	9.112	10.211	15.362	12.124	13.214
Proposed model	13.723	10.086	9.595	14.941	13.412	14.616

3.6   Conclusion

The principal applications of a convolutional neural network (CNN), which comprises one or even more convolutional layers, are preprocessing, categorization, segmentation, and other auto-correlated data.

In this work, the performance of the CNN, a popular image processing technique, in time series analysis will be improved. Based on the analysis' findings, it can be said that the CNN with the ideal smoothing factor outperforms other chosen time series forecasting techniques. In this study's optimal alpha, the finest assessment outcomes are obtained. Due to the created golden ratio, using Lucas numbers as hidden layers considerably improves the performance of the forecasting method.

Despite the fact that the outcomes addressed the research goals, this study still has certain restrictions. The implementation of efficient exponential smoothing in basic deep learning techniques is the study's main objective. As a result, future research will examine the impact of applying this strategy to more sophisticated deep learning algorithms (such as Resnet and hybrid CNN-LSTM). The investigation of various smoothing methods for trend data and the use of double- or triple-exponential smoothing are the topics of our following section. Further study will also take into account the utilization of multivariate data.

A CNN-based time-series forecasting method for the stock index has been developed in this work; the suggested model is utilized to provide prediction rules. The Figures 3.3–3.20 are shown for BSE SENEX, TAIEX, and KOSPI, respectively. The comparison results are shown in Tables 3.3–3.6 respectively. The outcomes showed that the suggested model performed better in terms of forecasting. Though 2020 is the COVID-19 period, other research papers have not tried to forecast this year as the stock index was fluctuating significantly, but we have tried to apply our method in 2020, and the forecasted outcome has shown satisfactory result. Tables 3.1 and 3.2 shows that Relu activation function with 100 epochs and five layers gives the best outcome.

The prediction method is easy to use, intuitive, and easy to describe the components of the model. In addition, field knowledge can be input into the model, for example, through specific points of change or power constraints. Prediction is not bad, but in some cases, you need to tweak certain parameters from their default values, so it is easy to do.

Acknowledgement

I need to thank Shri Partha Pratim Deb for his assistance for the duration of this work. I moreover want to thank the analysts for introducing positive grievance and ideas. Their knowledge and grievance significantly display the contemplations communicated on this paper.

References

Anuradha, J. (2021). Big data based stock trend prediction using deep cnn with reinforcement-lstm model. *International Journal of Systems Assurance Engineering and Management* 1–11. https://doi.org/10.1007/s13198-021-01074-2.

Chen, Y.C. and Huang, W.C. (2021). Constructing a stock-price forecast CNN model with gold and crude oil indicators. *Applied Soft Computing* 112: 107760.

Chen, W., Jiang, M., Zhang, W.G., and Chen, Z. (2021). A novel graph convolutional feature based convolutional neural network for stock trend prediction. *Information Sciences* 556: 67–94.

Chen, S.M. (1996). Forecasting enrollments based on fuzzy time series. *Fuzzy sets and systems* 81 (3): 311–319.

Durairaj, D.M. and Mohan, B.K. (2022). A convolutional neural network based approach to financial time series prediction. *Neural Computing and Applications* 34 (16): 13319–13337.

Dwivedi, S.A., Attry, A., Parekh, D., and Singla, K. (2021). Analysis and forecasting of Time-Series data using S-ARIMA, CNN and LSTM. *2021 International Conference on Computing, Communication, and Intelligent Systems (ICCCIS)*, Greater Noida (19–20 February 2021), pp. 131–136). IEEE.

Huarng, K.-H., Tiffany, H.-K.Y., Yu, W.H. (2007). A multivariate heuristic model for fuzzy time-series forecasting. *IEEE Transactions on Systems, Man, and Cybernetics, Part B (Cybernetics)* 37 (4): 836–846.

Lu, W., Li, J., Wang, J. et al. (2021). A CNN-BiLSTM-AM method for stock price prediction. *Neural Computing and Applications* 33: 4741–4753.

Mukherjee, S., Sadhukhan, B., Sarkar, N. et al. (2021). *Stock Market Prediction Using Deep Learning Algorithms*. CAAI Transactions on Intelligence Technology.

Nti, I.K., Adekoya, A.F., and Weyori, B.A. (2021). A novel multi-source information-fusion predictive framework based on deep neural networks for accuracy enhancement in stock market prediction. *Journal of Big data* 8 (1): 1–28.

Wang, H., Wang, J., Cao, L. et al. (2021). A stock closing price prediction model based on CNN-BiSLSTM. *Complexity 2021*: 1–12.

Ye, R. and Dai, Q. (2021). Implementing transfer learning across different datasets for time series forecasting. *Pattern Recognition* 109: 107617.

Yu, T.H.-K, and Huarng, K.-H. (2008). A bivariate fuzzy time series model to forecast the TAIEX. *Expert Systems with Applications* 34 (4): 2945–2952.

Zhang, X., He, K., and Bao, Y. (2021). Error-feedback stochastic modeling strategy for time series forecasting with convolutional neural networks. *Neurocomputing* 459: 234–248.

4

Comparative Study for Applicability of Color Histograms for CBIR Used for Crop Leaf Disease Detection

Jayamala Kumar Patil[1], Sampada Abhijit Dhole[2], Vinay Sampatrao Mandlik[3], and Sachin B. Jadhav[4]

[1] Department of Electronics and Telecommunication Engineering, Bharati Vidyapeeth's College of Engineering, Kolhapur, Shivaji University Kolhapur, Maharashtra, India
[2] Department of Electronics and Telecommunication Engineering, Bharati Vidyapeeth's College of Engineering for Women, Pune, Savitribai Phule Pune University, Maharashtra, India
[3] Department of Electronics and Telecommunication Engineering, Bharati Vidyapeeth's College of Engineering, Kolhapur, Shivaji University Kolhapur, Maharashtra, India
[4] School of Computer and System Sciences, Jawaharlal Nehru University (JNU) New Delhi, India

4.1 Introduction

In today's era of Industry Revolution 4.0, farming is also smarter, which includes automation in watering, fertilization, and harvesting with advanced technology in instruments and sensors using IoT. Yet there is a need for advanced algorithms for fast and accurate detection of diseases in crops.

In the case of crops, the disease is defined as any destruction of the usual physiological function of crops that generates some distinguishing symptoms, which are treated as proof of its existence. These are generated by one or more pathogens on crop leaves, stems, flowers, fruits, and roots. The disease also produces different characteristics and traits, like changes in the size, shape, and appearance of leaves, stems, flowers, and fruits. It changes green leaves, showing symptoms of it in terms of change in color, shape (shrinked), and rough texture.

The green contents of the leaf reduce as the disease spreads on it. It affects the photosynthesis activity of plants, which in turn reduces the food-forming process and hence overall productivity. Apart from this, pesticides used to control disease damage the food chain, may produce secondary pests, pose a human health risk,

Machine Learning Applications: From Computer Vision to Robotics, First Edition.
Edited by Indranath Chatterjee and Sheetal Zalte.
© 2024 The Institute of Electrical and Electronics Engineers, Inc.
Published 2024 by John Wiley & Sons, Inc.

and may result in acute and chronic health problems. Early diagnosis of disease enables timely treatment and may reduce the quantity of pesticides used. Agricultural experts may not be an efficient source for disease detection as humans suffer from subjectivity and inconsistency in their work.

This leads to the development of computer-based and customized systems for disease detection and expertization about treatment at any stage and any time based on computer vision technologies. These are intelligent computer programs that are competent in providing advice on specific problems in a given field, in a way and at a level of a human expert in that field. These systems work with reduced information and perform tasks more consistently than human experts (Maliappis et al. 2008). Such a system works on the principle of image pattern understanding and recognition. It identifies disease by comparing the input image of the diseased part of the crop with stored images. This mechanism resembles that of a human expert who uses knowledge and experience to detect the disease by comparing the plant organ under study with the knowledge database about plants stored in his/her brain. Content-based image retrieval (CBIR) systems work on the same principle of retrieving similar images from a stored database. Hence, in this paper CBIR system is developed to detect the leaf disease described by color and accurately retrieve similar ones in a faster way.

4.2   Literature Review

Extensive literature is available on the development of CBIR systems using various image features. The literature related to the detection of crop/plant diseases is discussed in this section.

Mohanty et al. (2016) showed that the difficult task of rapid identification of crop diseases is possible by smartphone-assisted disease diagnosis with computer vision by deep learning. A deep CNN was trained using diseased and healthy plant leaves collected under controlled environment. Fourteen crops and twenty-six diseases were examined. The developed model succeeded in obtaining an accuracy of 99.35%, indicating a path toward smartphone-assisted crop disease diagnosis.

Singh and Mishra (2017) presented a review of disease classification techniques and image segmentation techniques that can be used for the automatic detection and classification of leaf diseases later. Different plant species including Banana, beans, jackfruit, and lemon, are used and their diseases are identified. The authors further noted that to improve the recognition rate in the classification process Artificial Networks, Bayes classifiers, Fuzzy Logic, and hybrid algorithms can also be used.

A method based on different color models and Kapur's thresholding for plant disease detection is developed by Tuba et al. (2017). Four different color models, i.e. RGB, YCbCr, HSI, and CIELAB, were tested and compared. The authors noted that compared to other color models, HSI is more suitable for images with noise that is introduced due to background, vein, and camera flash and provides the best results. Further, it is observed that RGB is practically unusable. Component H was used for image segmentation, where diseases were separated from the leaf. A median filter was applied to color-transformed image. The disease spot area is determined by applying Kapur's threshold to different color components.

A research study by Jos and Venkatesh (2020) reports pseudo-HSV color region features. These are used for Tomato disease identification. The developed algorithm detects eccentric and non-eccentric dots, spots, patches, and regions of different pseudocolors to generate effective feature vectors.

Ahmad et al. (2021) proposed a model for the detection and identification of leaf diseases. It involves extraction of diseased leaf area and calculation of a set of 29 texture and color features. Feature standardization is used to reduce the bias due to feature range and feature selection and is performed to select the most discriminatory features. Of the various classifiers tested, which include SVM, K-NN, Naïve Bayes, random forest, and artificial neural networks, SVM provided the best performance over a dataset of 227 images. The classification accuracy of diseased and healthy plant identification is 91.40%, and for disease category identification is 82.47%. The authors reported that the only limitation is that the system will be able to classify only images that are visually distinguishable and provided in the training database.

Sharad Hasan et al. (2022) proposed a machine learning and computer vision-based automated system using $L^*a^*b^*$ color space to segment disease-infected regions of apple disease. Derived color markers are based on a^*b^* color space. DWT features of segmented images are extracted using a multilevel discrete wavelet transformation on each row of the segmented image. The fusion of DWT features and $L^*a^*b^*$ color histogram achieved a predictive accuracy of 98.63%.

The study of all the above research papers used multiple features, and a neural network-controlled environment for image acquisition, which is computationally complex and time-consuming. Ahmad et al. (2021) indicated that only trained images are classified by the designed system. These limitations suggest the need for the development of systems that are computationally efficient and will detect and identify diseases based on real-time images obtained in controlled as well as uncontrolled environments by any person/farmer at any time by digital camera or mobile camera. Such a system based on CBIR is proposed in this paper.

4.3 Methodology

A CBIR-based system for retrieving diseased leaves from crops to detect and recognize the disease is developed. The heart of the CBIR system is the extraction of relevant image features, Feature Matching, and an image database, as shown in Figure 4.1.

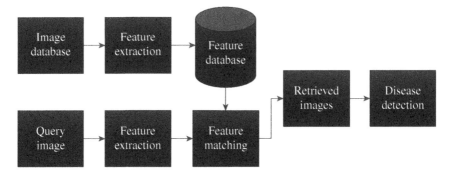

Figure 4.1 CBIR system for disease detection.

Feature extraction converts images from high- to low-dimension feature space descriptions. Image feature produces some quantifiable characteristics of an object as a function of one or more measurements. The features are broadly classified into low-level features, which can be derived from the original image, and High-Level features, which can be derived from low-level features (Ryszard 2007). Features are also classified as general features, which do not depend on the application (e.g. color), and domain specific features, which depend on the application and can be computed for the whole image or part of an image. Feature matching involves the comparison of the feature vector of the query image with the feature vectors of each image in the database. This comparison provides a means of defining a threshold value for the CBIR system to retrieve similar images. The decision of whether a leaf is healthy or diseased is taken on the precision of retrieved images.

In this research a developed CBIR system is implemented using color features of leaves, as color provides significant information about the disease. The color of a diseased leaf differs from a healthy leaf. It is subject to disease type. e. g. soybean leaf affected by Septoria brown spot (SBS) disease looks reddish, whereas healthy leaf appears green.

4.3.1 Color Features

The human visual system is more sensitive to color images than gray images, which makes color as the most significant and powerful descriptor for object recognition. Compared to texture or shape features of images, color features are simple to extract, relatively vigorous to background conditions, and do not depend on the dimension and direction of the image (Mangijao Singh and Hemachandran 2012).

For color image processing, the color model/space play a significant role. Several color models are available. Section 4.3.1.1 – section 4.3.1.3 presents an overview of popular color models. In any color model, image color features can be obtained by various methods like color histograms (Ryszard 2007; Suhasini et al. 2009), discussed in detail in Section 4.3.1.4, correlogram (Huang et al. 1997), moments (Maheswary and Srivastav 2008), color structure descriptors, etc.

4.3.1.1 RGB Color Model/Space

This color model is reliant on the device. It uses Cartesian coordinate system. It is a cube with Red, Green, and Blue primary values at three corners and secondary colors cyan, magenta, and yellow at three other corners (Rafael and Richard 2008).

The space spanned by the R, G, and B describes visible colors. In RGB, any color is a point on or inside the cube, which is represented as a vector extending from the origin in the 3-D RGB color space. This makes it useful for representing the color features of images. On the other hand, the RGB model is perceptually nonuniform. It can be changed to generate other color spaces that are perceptually uniform.

4.3.1.2 HSV Color Space

HSV stands for hue (H), saturation(S), and value (V). It is also called as his, where I stand for intensity. Hue is an attribute of color that specifies a pure color, e.g. red, green, etc. Saturation is a measure of the dilution of pure color by white light. Brightness embodies an achromatic impression of Intensity. This color space separates intensity components from chromatic components in the distribution of colors in an image, which makes it an ideal means for its application in color image descriptions. Another fact is that the components of this color space closely correspond to human color perception because humans depict color items by hue, saturation, and brightness.

HSV color space is a nonlinear conversion of the RGB space. It is approximately perceptually uniform. It is extensively applicable to color vision. The HSV components are obtained from RGB using the following equations (Rafael and Richard 2008).

$$H = \cos^{-1}\left\{ \frac{\frac{1}{2}\left[(R-G)+(R-B)\right]}{\sqrt{\left[(R-G)^2+(R-B)(G-B)\right]}} \right\} \tag{4.1}$$

$$S = 1 - \frac{3\left[\min(R,G,B)\right]}{(R+G+B)} \tag{4.2}$$

$$V = \frac{(R+G+B)}{3} \tag{4.3}$$

4.3.1.3 YCbCr Color Space

Humans are comparatively more responsive to luminance than to chrominance. YCbCr color space takes advantage of this fact to attain proficient depiction of images. It decouples the luminance and chrominance components of an image (Roman 10 2011). Here luminance information is stored in "Y" component and chrominance information is stored in two color difference components "Cb" and "Cr." chrominance signal "Cb" is the difference between the blue component and the reference value while the chrominance signal "Cr" is the difference between the red component and the reference value. Compared to RGB, YCbCr color space is luminance independent, hence it gives better performance. It is generated from RGB space using the following equations (Charles Poynton 1995):

$$Y = 0.299(R - G) + G + 0.114(B - G) \tag{4.4}$$

$$Cb = 0.564(B - Y) \tag{4.5}$$

$$Cr = 0.713(R - Y) \tag{4.6}$$

4.3.1.4 Color Histogram

Histogram is a graphical illustration of the number of pixels present at various light intensities of an image. For a digital image, the histogram is (Rafael and Richard 2008) given by equation (4.7).

$$p(r_k) = \frac{n_k}{N} \tag{4.7}$$

where, r_k is the kth gray level, n_k is the number of pixels in the image with that gray level, and N is the total number of pixels in the image.

Similarly, a color histogram is a graphical illustration of the distribution of colors in an image that provides a statistic for an estimate of underlying color distribution. It is composed of three histograms corresponding to three color channels, R/G/B or H/S/V or Y/Cb/Cr, as shown in Figure 4.2.

Histograms are very easy to compute and insensitive to small variations in the images. It is relatively invariant to translation and rotation about the viewing axis. The comparison of the histograms of two images with suitable similarity measures like Euclidean distance, histogram intersection, correlation, cosine, or quadratic distances provides a suitable platform for classifying or recognizing an object.

4.3.2 Database

This research work is designed to detect various diseases of soybean leaf. As no ready-made and standard database is available, leaf images are collected from

(a)

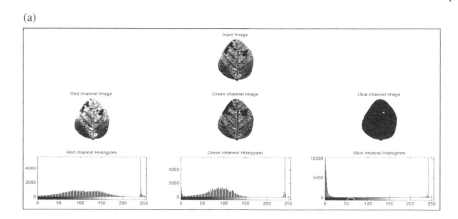

(b)

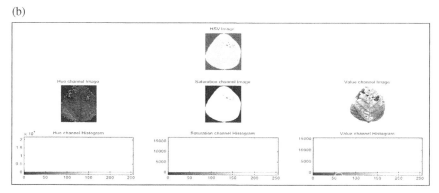

(c)

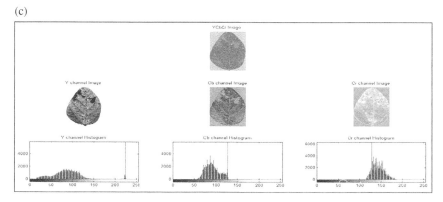

Figure 4.2 Color histogram in RGB, HSV, and YCbCr color spaces. (a) RGB histogram, (b) HSV histogram, (c) YCbCr histogram.

various fields in Kolhapur and Sangli districts of the state of Maharashtra, India. These images were captured by a digital camera with 16 megapixels on a white background.

Soybean is the main oil seed crop in India. Madhya Pradesh is known as the soya state of India. It produces about 55% of total production, followed by Maharashtra (30%). Rajasthan produces about 9%, Karnataka contributes 2%, Chhattisgarh and Andhra Pradesh contribute 1% each, and other states contribute 2% of the total production. In India, about 35 diseases have been identified in soybean (Wrather et al. 2010). A database of soybean is created from samples of leaves of two soybean diseases and healthy leaves. Soybean has basically elliptic-shaped leaves. The details of Soybean leaf diseases taken for study, their symptoms (Cruz et al. 2010), their effect on production, and the number of samples used for training and testing are summarized in Table 4.1.

Table 4.1 Soybean diseases.

| | | | No of samples of leaf | |
| | | | Feature database | Query database |
Disease	Symptoms	Effects/losses		
Mosaic virus	**Pathogen**: Alfalfa Mosaic Virus 1) Symptoms vary from mosaic to mottle patterns in the form of bright yellow and dark green leaf tissues 2) Young leaves show light yellow spots and vein browning. 3) Localized dead lesions on leaflets. 4) Bright yellow mosaic on leaves; or leaf veins are yellow but the remainder of the leaf is a normal green color.	1) Stunted plants produce few pods. 2) Yield loss of 32–48%.	100	155

Table 4.1 (Continued)

Disease	Symptoms	Effects/losses	No of samples of leaf	
			Feature database	Query database
Septoria brown spot	**Pathogen**: Septoria Glycines 1) Small, irregularly shaped brown to red–brown spots or lesions on the upper and lower leaf surfaces. 2) Diseased leaves become tarnished brown or yellow and fall ahead of time. 3) May merge into large areas of dead leaf tissue.	1) Results in smaller seed size. 2) Yield losses of 12–17% may occur.	100	111
Healthy	Green leaves	—	100	110
		Total samples **300**		**376**

4.3.3 Parameters for Performance Analysis

The performance of the CBIR system is evaluated using precision and recall as stated in equations (4.8) and (4.9), respectively. Precision provides a measure of exactness, while recall provides a measure of completeness (Wang and Qin 2009).

$$\text{Retrieval Precision}(p) = \frac{\text{number of relevant images retrieved}(Ri)}{\text{total number of images retrieved}(Tri)} \tag{4.8}$$

$$\text{Recall}(r) = \frac{\text{number of relevant images retrieved}(Ri)}{\text{total number of relevant images}(Tr)} \tag{4.9}$$

The disease detection efficiency of each method is computed using equation (4.10). It is the ratio of correctly disease-detected leaf images by total leaf images tested during experimentation.

$$\text{Disease Detection Efficiency}(e) = \frac{Ic}{It} \times 100 \qquad (4.10)$$

where "Ic" is the number of correctly detected images, and "It" is the total number of tested images.

4.3.4 Experimental Procedure for CBIR Using Color Histogram for Detection of Disease

Figure 4.3a shows the procedure used to develop feature database. It consists of image preprocessing in which input images are resized to web-small size. Then the image is converted either into HSV or YCbCr color space, or it is used as it

(a)

(b)

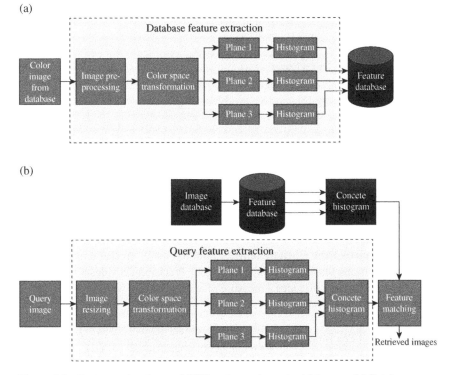

Figure 4.3 Feature extraction and CBIR system using color histogram. (a) Database feature extraction using color histogram, (b) CBIR system using color histogram.

is, i.e. in RGB color space. Each of these color spaces will have three color planes. The histogram is obtained for each plane and stored as a color feature. The experimental procedure used to retrieve diseased images is shown in Figure 4.3b.

Algorithm used to formulate feature database is stated in algorithm 4.3.1 as below:

When a query image is given, its features are extracted in a similar manner to the extraction of features in database images. The three plane histograms are concatenated to form a single query feature vector. The size of this feature vector is (768X3), as the size of one plane histogram is 256 bins. These are compared with the similar feature vector of each image in the database. The images that show similarity are extracted as the best-matched images. For similarity measurement, performance using Euclidian distance as well as the correlation coefficient between query and database image histograms is evaluated, which are given by equations (4.11) and (4.12), respectively. It is found that the correlation coefficient provides better results as compared to distance matching.

$$\text{Dist}(Q,D) = \sum_{i=0}^{l-1}(Q_i - D_i) \tag{4.11}$$

where, Q_i is the feature vector of the query image, D_i is the feature vector of the database images, and l is the dimension of the feature.

$$\text{correlation coefficient}(r) = \frac{\sum_m \sum_n (Q_{mn} - Q')(D_{mn} - D')}{\sqrt{\sum_m \sum_n (Q_{mn} - Q')^2 \left(\sum_m \sum_n (D_{mn} - D')^2\right)}} \tag{4.12}$$

where Q' is the mean of the query image histogram and D' is the mean of the database image histogram.

The algorithm used for development of CBIR using color histogram is in algorithm 4.3.2 stated as:

Algorithm 4.3.1   Creation of Feature Database Using Color Histogram

Step 1: Input a color leaf image.
Step 2: Preprocess the image to resize it.
Step 3: Convert the image into another color space if applicable (e.g. HSV or YCbCr).
Step 4: Find the histogram of each plane and store it as a color feature.

Algorithm 4.3.2 CBIR Using Color Histogram
Step 1: Input query color leaf image. Step 2: Preprocess the image to resize it. Step 3: Convert the image into another color space if applicable (e.g. HSV or YCbCr). Step 4: Find the histogram of each plane. Step 5: Concate all three histograms to produce query color histogram, HQ. Step 6: Similarly, concate all three histograms of each database image by using three color plane histograms from feature database HD. Step 7: For similarity measurement find the correlation coefficient between HQ and HD. Step 8: Arrange these values in descending order and retrieve similar images. Step 9: The disease of the query image is the same as that of the disease of the maximum retrieved images.

4.4 Results and Discussions

This section evaluates results obtained by implementing algorithms discussed in Section 4.3.5 for two diseases of soybean, i.e. mosaic and SBS, as well as healthy leaf. The experiment is repeated several times in RGB, HSV, and YCbCr color space, and results are evaluated for randomly selected ten images from each disease and healthy category. The performance is evaluated using retrieval precision, recall, and disease detection efficiency. The top 40 retrieved images are observed, and the decision of disease detection is based on the precision of the top 10 images.

4.4.1 Results of CBIR Using Color Histogram for Detection of Soybean Alfalfa Mosaic Virus Disease

As shown in Figure 4.4a, the query image is a soybean leaf affected by alfalfa mosaic virus (AMV) disease. This retrieval using RGB histogram shows that similar leaves are extracted at the 3rd, 5th, and 9th positions in the top 10 retrievals. As for the detection precision of the top 10 retrievals is considered, the RGB color model failed to detect the disease. It has detected the query diseased leaf as a healthy leaf, as most of the images retrieved are of healthy class. Figure 4.4b shows that the HSV color model successfully retrieved all 10 similar images in the top 10 retrievals, detecting disease correctly. In fact, compared with the RGB color model, the overall performance of the HSV color model is improved in the top 40

(a)

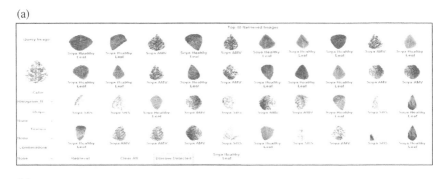

(b)

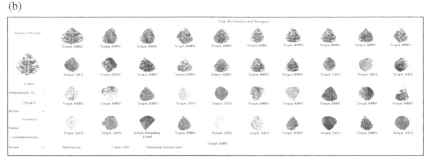

(c)

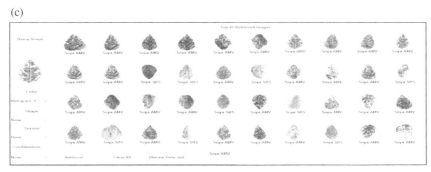

Figure 4.4 Detection of soybean alfalfa mosaic virus (AMV) disease using RGB, HSV, and YCbCr histograms. (a) CBIR RGB histogram, (b) CBIR using HSV histogram, (c) CBIR using YCbCr histogram.

retrievals as well. Figure 4.4c shows the retrieval using YCbCr color space, indicating its success in detecting the disease and retrieving more relevant images in top 10 as well as top 40 positions compared to HSV as well as the RGB color model. Though in the top 10 retrievals, the performance of HSV and YCbCr is equivalent, YCbCr has retrieved more relevant images, proving its ability as a significant feature for CBIR.

4.4.2 Results of CBIR Using Color Histogram for Detection of Soybean Septoria Brown Spot (SBS) Disease

Here query image is a soybean leaf affected by SBS disease. As shown in Figure 4.5a, the RGB histogram correctly detected disease with a precision of 80% in the top 10 retrievals and 100% in the top 5 retrievals. Figure 4.5b shows retrieval

(a)

(b)

(c)

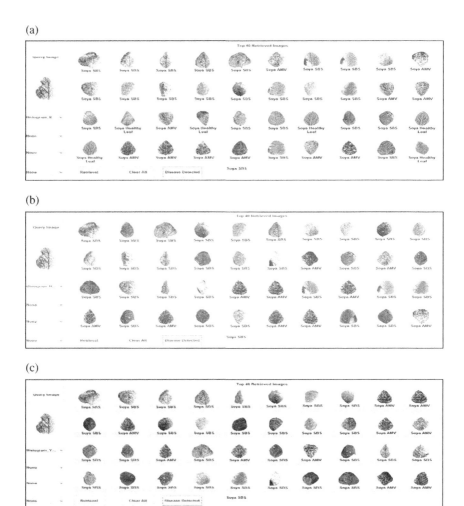

Figure 4.5 Detection of soybean Septoria Brown Spot (SBS) disease using RGB, HSV, and YCbCr histograms. (a) CBIR using RGB histogram, (b) CBIR using HSV histogram, (c) CBIR using YCbCr histogram.

using an HSV histogram having 100% precision in the top 5 and top 10 retrievals and correct disease detection. Compared to the RGB color model, the result is improved in terms of the quality of retrieved images for the HSV color model. The performance of the YCbCr color model for SBS is shown in Figure 4.5c. It was found that the top 5 retrievals gave 100% precision, and the top 10 provided 80% precision with correct disease detection. This performance is equivalent to the RGB histogram method, but in RGB color space, AMV is retrieved at the 6th position, while in YCbCr it is retrieved at the 9th position, showing its improvement over RGB. Compared to the HSV histogram method, the YCbCr histogram method has lower performance, but the quality of images retrieved is better for a given query image.

4.4.3 Results of CBIR Using Color Histogram for Detection of Soybean Healthy Leaf

A healthy leaf is correctly detected by all methods in this research, as shown in Figure 4.6.

Table 4.2 shows the average precision, recall, and disease detection efficiency of all color models tested in this research computed for randomly selected 10 images of all leaf types for top 10 retrievals. The average disease detection efficiency of 96.66% is provided by CBIR system using YCbCr Histogram.

4.5 Conclusion

The color of an image is one of its significant features. Several researchers extracted various color features of an image for the purpose of image classification, recognition, detection, and retrieval. Histograms and various moments of an image are the most favorable features for the purposes of image detection, classification, and retrieval. These color features are extracted in three color spaces: RGB, HSV, and YCbCr. Many researchers have experimented with histogram and moment features in these color spaces for various applications such as image categorization, retrieval, and detection of diseases of leaves, fruits, etc. Research in this paper deals with the testing of color histograms for the development of the CBIR system. The applicability of this system is tested for the purpose of detection of soybean leaf diseases. Three leaf types of soybean, which include two disease classes along with healthy leaves are tested, but there are still a lot of disease classes that need attention. It was found that of the tested RGB, HSV, and YCbCr color spaces, the histogram in YCbCr color space resulted in the

(a)

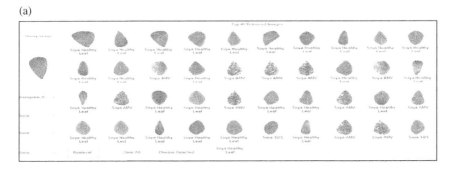

(b)

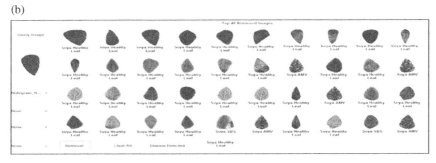

(c)

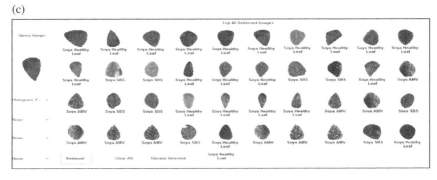

Figure 4.6 Detection of soybean Healthy leaf using RGB, HSV, and YCbCr histograms. (a) CBIR using RGB histogram, (b) CBIR using HSV histogram, (c) CBIR using YCbCr histogram.

best performance and provided 96.66% disease detection for the images collected in an uncontrolled environment. In the future, it can be integrated with additional texture and/or shape features to provide global solutions for all leaf disease detection problems.

Table 4.2 Average precision, recall, and disease detection efficiency of CBIR using YCbCr histogram.

Leaf type	Color space/ model	Top 10 Retrievals		Disease detection efficiency (%)	Average disease detection efficiency (%)
		Precision	Recall		
Soybean Alfalfa Mosaic Virus	RGB	0.62	0.062	60	80
Soybean Septoria Brown Spot		0.50	0.050	80	
Soybean healthy leaf		0.73	0.073	100	
Soybean Alfalfa Mosaic Virus	HSV	0.82	0.082	80	90
Soybean Septoria Brown Spot		0.78	0.078	90	
Soybean healthy leaf		0.80	0.080	100	
Soybean Alfalfa Mosaic Virus	YCbCr	0.86	0.086	100	96.66
Soybean Septoria Brown Spot		0.82	0.082	90	
Soybean healthy leaf		0.87	0.087	100	

References

Ahmad, N., Asif, H.M.S., and Saleem, G. (2021). Leaf image-based plant disease identification using color and texture features. *ArXiv* 2102: 04515.

Aruna, P., Vandana, M., Mrudula, M. (2020). Skin Segmentation of Indian Sign Language Recognition System for Differently- Able People, *International Journal of Advances in Science Engineering and Technology* 8 (2): 6–11.

Cruz, C.D., Mills, D., Paul, P.A., and Dorrance, A.E. (2010). Impact of brown spot caused by *Septoria glycines* on soybean in ohio. *The American Phytopathological Society, Plant Disease* 94 (7): 820–826.

Hasan, S., Sarwar Jahan, M., and Islam, I. (2022). Disease detection of apple leaf with combination of color segmentation and modified DWT. *Journal of King Saud University – Computer and Information Sciences* 34 (9): 7212–7224. https://doi.org/10.1016/j.jksuci.2022.07.004.

Huang, J., Kumar, S.R., Mitra, M. et al. (1997). Image indexing using color correlograms. *IEEE Int Conf Computer Vision and Pattern Recognition,* San Juan, Puerto Rico (17–19 June 1997), pp.762–768. IEEE.

Jos, J. and Venkatesh, K.A. (2020). Pseudo color region features for plant disease detection. *2020 IEEE International Conference for Innovation in Technology (INOCON)*, Bangluru (6–8 November 2020), pp. 1–5. https://doi.org/10.1109/INOCON50539.2020.9298337.

Maheswary, P. and Srivastav, N. (2008). Retrieving similar image using color moment feature detector and K-means clustering of remote sensing images. *International Conference on Computer and Electrical Engineering,* Phuket (20–22 December 2008), pp. 821–824. IEEE.

Maliappis, M.T., Ferentinos, K.P., Passam, H.C., and Sideridis, A.B. (2008). Gims: a web based greenhouse intelligent management system. *World Journal of Agricultural Sciences* 4 (5): 640–647.

Mangijao Singh, S. and Hemachandran, K. (2012). Content based image retrieval using color moment and gabor texture feature. *International Journal of Computer Science* 9 (5): 299–309.

Mohanty, S.P., Hughes, D.P., and Salathé, M. (2016). Using deep learning for image-based plant disease detection. *Frontiers in Plant Science* 7: 1419. https://doi.org/10.3389/fpls.2016.01419.

Charles Poynton (1995). Color spaces. http://www.compression.ru/download/articles/color_space/ch03.pd.

Rafael, C.G. and Richard, E.W. (2008). *Color Image Processing*, 3e. Pearson Education Inc.

Roman 10 (2011). YCbCr color space – an intro and its applications. *A Journey to Software Craftsmanship*. http://www.roman10.net/2011/08/18/ycbcr-color-spacean-intro-and-its_applications/.

Ryszard, S.C. (2007). Image feature extraction techniques and their applications for CBIR and biometrics systems. *International Journal of Biology and Biomedical Engineering* 1 (1): 6–16.

Singh, V. and Mishra, A.K. (2017). Detection of plant leaf diseases using image segmentation and soft computing techniques. *Information Processing in Agriculture*. Elsevier 4 (1): 41–49.

Suhasini, P.S., Krishna, K.S.R., and Krishna, I.V.M. (2009). CBIR using color histogram processing. *Journal of Theoretical and Applied Information Technology* 6 (1): 116–122.

Tuba, E., Jovanovic, R., and Tuba, M. (2017). Plant diseases detection based on color features and kapur's method. *WSEAS Transactions on Information Science and Applications* 14: 31–39.

Wang, S. and Qin, H. (2009). A study of order- based block color feature image retrieval compared with cumulative color histogram method. *Sixth International Conference On Fuzzy Systems and Knowledge Discovery, IEEE Computer Society*, Tianjin (14–16 August 2009), pp.81–84. IEEE.

Wrather, A., Shannon, G., Balardin, R. et al. (2010). Effect of diseases on soybean yield in the top eight producing countries in 2006. *Plant Health Progress* https://doi.org/10.1094/PHP-2010-0125-01-RS.

Biographies of Authors

Dr. Jayamala Kumar Patil holds a Doctor of Electronics Engineering degree from Bharati Vidyapeeth (Deemed to be) University, Pune, India. She received her Batcher and Master of Engineering in Electronics from Shivaji University, Kolhapur, India in 1999 and 2008, respectively. She is currently an associate professor at Department of Electronics and Telecommunication Engineering in Bharati Vidyapeeth's College of Engineering, Kolhapur, Maharashtra, India. Her research includes signal processing, image processing, artificial intelligence. She has published over 60 papers in international journals and conferences. She can be contacted at email: jayamala.p@rediffmail.com

Dr. Sampada Abhijit Dhole has completed Ph.D. in Electronics from Bharati Vidyapeeth (Deemed to be university) College of Engineering, India in 2017 with specialization of Image Processing and Biometrics. Her research interest includes the Image processing, multimodal. She has published more than 30 research papers. She is working as assistant professor in the department of E&TC at Bharati Vidyapeeth's college of Engineering for Women, SPPU, Pune, India. She has 19 years of teaching experience. She is a member of technical society ISTE, India. She can be contacted at email: sampada.dhole@bharatividyapeeth.edu

Vinay Sampatrao Mandlik holds a Bachelor of Engineering (B.Eng.) in Electronics and Telecommunication Engineering, Master of Technology (M. Tech) in Electronics Engineering form Shivaji University of Kolhapur, India. He is currently Assistant Professor at department of Electronics and Telecommunication Engineering in Bharati Vidyapeeth's College of Engineering, Kolhapur, Maharashtra, India. His research areas of interest include Image Processing and Artificial Intelligent. He has published over 10 papers in international journals and conferences. He can be contacted at email: vinaymandlik@gmail.com

Dr. Sachin Jadhav holds a Doctor of Electronics Engineering degree from VTU Karnataka, India. He received his Batcher and Master of Engineering in Electronics from Shivaji University, Kolhapur, India. He is currently associate professor at School of Computer & Systems Sciences (SC&SS, Jawaharlal Neharu University (JNU) New Delhi-110067, India. His research includes signal processing, image processing, Artificial Intelligence. He has published papers in international journals and conferences. He can be contacted at email: sbjadhav@bvucoep.edu.in

5

Stock Index Forecasting Using RNN-Long Short-Term Memory

Partha Pratim Deb[1], Diptendu Bhattacharya[1], and Sheetal Zalte[2]

[1] *Department of Computer Science and Engineering, National Institute of Technology Agartala, Agartala, Tripura, India*
[2] *Department of Computer Science, Shivaji University, Kolhapur, Maharashtra, India*

5.1 Introduction

Scholars and researchers from numerous academic backgrounds have been drawn to the issue of financial market forecasting. The efficient market hypothesis, first proposed by Eugene Fama in the 1960s, states that financial markets are informationally efficient, meaning that stock prices fully reflect all available information at any given time. This hypothesis has been the subject of much debate over the years, with some arguing that markets are indeed efficient, while others argue that there are persistent inefficiencies that can be exploited through careful analysis and modeling.

Regardless of whether or not the efficient market hypothesis holds true, machine learning techniques can be useful for identifying patterns and trends in historical stock price data that may not be immediately apparent to human analysts. However, it is important to keep in mind that stock price prediction is a complex and uncertain task, and any predictions made using machine learning models should be treated with caution and used in conjunction with other forms of analysis and expert knowledge. The hypothesis of Burton et al. states that predicting or forecasting the financial market is impractical due to the unpredictability of price movements in the actual world (Malkiel 1989).

The goal of this research is to figure out what time period and how many epochs are ideal for predicting certain stock values. We concentrated on three equities

Machine Learning Applications: From Computer Vision to Robotics, First Edition.
Edited by Indranath Chatterjee and Sheetal Zalte.
© 2024 The Institute of Electrical and Electronics Engineers, Inc.
Published 2024 by John Wiley & Sons, Inc.

with high RMSE values: TAIEX, BSE, and KOSPI. We gathered 5 years of actual and predicted BSE, KOSPI and TAIEX data and utilized it to train and validate the model.

Such stocks were picked specifically to see how an RNN would handle titles from the same subindustry, where two of them are projected to be highly connected and the remaining one is near to disrupting a competitor/innovator for more than two economic cycles.

Saratha Sathasivam proposed the Hopfield network in 1982 (Hopfield 1982), from which the RNN was derived. Due to the traditional machine learning algorithm's heavy reliance on human feature extraction, image recognition, audio recognition, and natural language processing experience feature extraction bottlenecks (Zheng and Zheng 2019; Wei et al. 2021). In comparison to other neural network models, the RNN is better at processing sequential data because it "remembers" the time series' sequential context (Ni and Cao 2020).

Figure 5.1 depicts the typical construction of an RNN, which allows data from the initial input to easily flow to the final output. However, when the duration of the series increases, the problem of diminishing gradients develops (Chung et al. 2014).

During training with gradient descent, the RNN backpropagates over time from the last hidden layer to the first hidden layer. Because a single hidden layer's partial derivative can only have an absolute value of 1 due to the activation function, multiplying a lengthy chain of partial derivatives leads the gradient to converge to 0. The training process is slowed by the diminishing gradient, which results in a strong bias toward the input data as time progresses. Long Short-Term Memory (LSTM) which was constructed by Cho (Cho et al. 2014) is presented as a solution to this problem.

The following sections of the article are organized as follows: The materials of this method are explained in Section 5.2. Section 5.3 discusses RNN-based stock

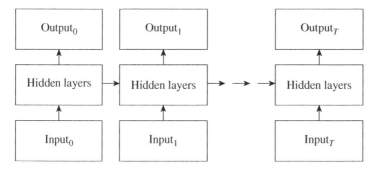

Figure 5.1 RNN structure.

market methodologies, in which this paper will go through each phase in depth.; Section 5.4 offers the results and discussion; and Section 5.5 concludes the survey and discusses the RNN's potential uses in stock market prediction.

5.2 Materials

We have considered three unique sorts of indices for our experimental purpose, where three facts are set, KOSPI, TAIEX and BSE-SENSEX. From the years 2015 to 2020, 365 days of records for each year are used for the inventory index. Data from January 1 to October 31 are used for training and those from November 1 to December 31 are used for testing for evaluating the effectiveness of the RNN.

5.3 Methodology

5.3.1 RNN

The analysis and forecasting of sequence data is the RNN's main duty. Figure 5.2 shows the model structure, with xt denoting the input of the training sample at time t in the time series. This represents the time-hidden series' state for the model at time t, the time series output of the model.

It is obvious that models exist that might serve as examples for both Deep Learning and Time Series Analysis in Machine Learning since they have evolved through various procedures. The method that is being suggested for this project relies around CNN models, which are excellent for forecasting. The hidden states $h_1, h_2, ..., h_t$ correspond to the inputs $x_1, x_2, ..., x_t$, respectively, while the outputs are

Figure 5.2 RNN model.

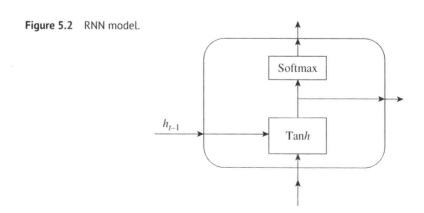

$y_1, y_2, \ldots, y_t$. The computation procedure of a conventional RNN, for example, may be written as formula (5.1) (5.2)

$$h_t = f(U x_t + W h_t - 1 + b) \tag{5.1}$$

$$y_t = \text{softmax}(V h_t + c) \tag{5.2}$$

U, W, and V stand for the weight parameters, b and c for the bias parameters, and f for the activation function, which is often the tan h function (). The RNN is susceptible to gradient disappearance and explosion problems when the data sequence is sufficiently long, despite the fact that it may theoretically be able to handle the training of sequence data successfully. The RNN cannot therefore be used directly in application fields including voice recognition, handwritten books, language processing, and the processing of natural languages (NLP).

5.3.2 LSTM

Text labeling, time series analysis, voice recognition, and speech recognition are just a few of the uses for LSTMs, which are a specific category of RNNs. The LSTM approach was proposed by Hochreiter and Schmidhuber (Hochreiter and Schmidhuber 1997). The LSTM has the ability to recall values from earlier stages of an RNN for future usage. The RNN's capacity to record past values and assign precedence in older data sequences to learning from newer ones is boosted by LSTM units. In the typical LSTM, the prior data trend can be memorized through certain doors and a memory line.

Figure 5.3 depicts the LSTM structure. The forget gate, input gate, and output gate make up the LSTM. The three gates work together to govern data storage and erasure (Gers et al. 2000, 2001).

$$f_t = \sigma(W f h \cdot h_t - 1 + W f x \cdot x_t + b f) \tag{5.3}$$

According to the formula (5.3), the forget gate f_t determines how much data from the previous moment may be kept in the current instant, where $h_t - 1$ stores the data from the previous time step t 1, performs a linear transformation by multiplying with the weight matrix, and the bias is represented by Wfh.bf, x_t is the input vector for the t-th time step or the t-th element of the input sequence x, and x_t is linearly transformed by multiplying with the weight matrix $W fx$. The Sigmoid activation function is used to determine the sum of these three pieces of data.

$$i_t = \sigma(W_{ih} \cdot h_t - 1 + W_{ix} \cdot x_t + b_i) \tag{5.4}$$

Figure 5.3 LSTM model.

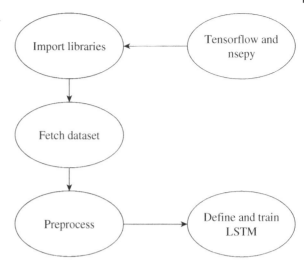

as indicated in the formula (5.4). The input gate regulates how much data information is maintained in the present instant.

$$o_t = \sigma\left(W_{oh} \cdot h_{t-1} + W_{ox} \cdot x_t + b_o\right) \tag{5.5}$$

$$c_t = f_t \times c_{t-1} + i_t \times \tanh\left(W_o \times [h_{t-1}, x_t] + b_o\right) \tag{5.6}$$

$$h_t = o_t \times \tanh(c_t) \tag{5.7}$$

as indicated in the formula (5.5). The output gate o_t regulates the amount of data that may be transmitted from one instant to the next. c_t is the data information kept from the beginning to the present moment, as illustrated in formula (5.6). h_t regulates how much data information from the beginning to the present instant may be transmitted to the next moment, as stated in formula (5.7).

5.4 Result and Discussion

The demonstration is executed utilizing the RNN with LSTM calculation. It is the 1-stacked layer show with the root implying square blunder values. The table shows the outcomes about of the datasets. The table is actualized utilizing the desire calculation. In this segment, this paper applies the proposed strategy to estimate the TAIEX, BSE, and KOSPI from 2015 to 2020. It shows the exploratory outcomes of the proposed strategy with one of the strategies displayed in Figures 5.4–5.6 represented the information for actual/forecasted results over the

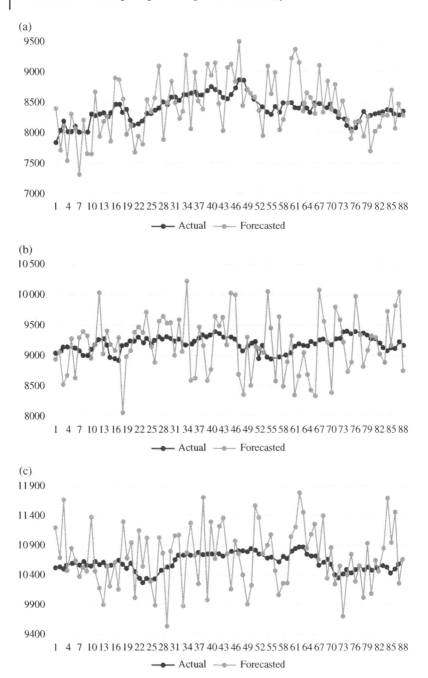

Figure 5.4 Actual vs forecasting for TAIEX from 2015 to 2020. The comparison tables show the above average of all the methods with year 2015 to 2020 for TAIEX, where (a) resembles 2015, (b) resembles 2016, (c) resembles 2017, (d) resembles 2018, (e) resembles 2019, and (f) resembles 2020

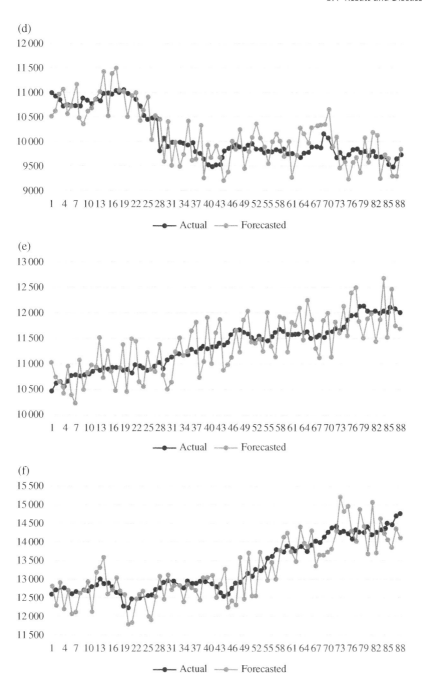

Figure 5.4 (Continued)

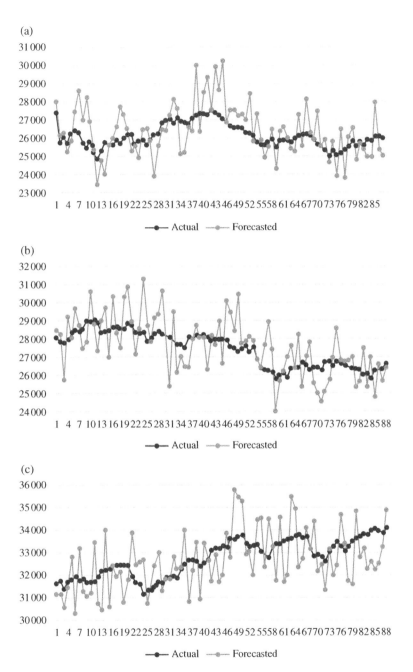

(a)

(b)

(c)

Figure 5.5 Actual vs forecasting for BSE-SENSEX from 2015 to 2020. The above graphical representation showed the data of BSE Limited which is also known as the Bombay Stock Exchange, which is an Indian stock exchange located on Dalal Street in Mumbai. Established in 1875 by cotton merchant Premchand Roychand, a Rajasthani Jain businessman, it is the oldest stock exchange in Asia and also the tenth oldest in the world. The comparison tables above show the above average of all the methods from the year 2015 to 2020 for BSE SENSEX, where (a) resembles 2015, (b) resembles 2016, (c) resembles 2017, (d) resembles 2018, (e) resembles 2019, and (f) resembles 2020.

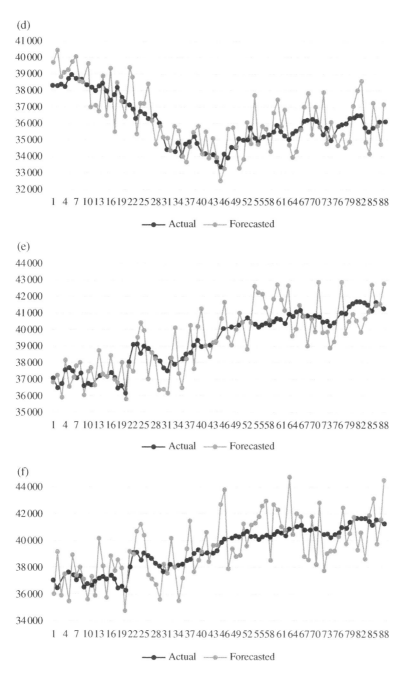

Figure 5.5 (Continued)

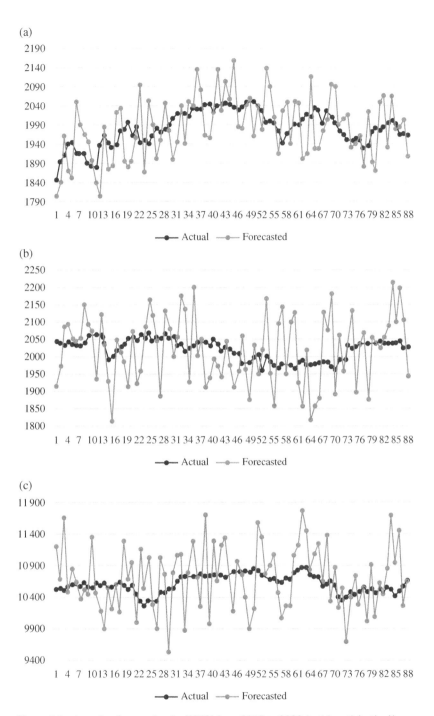

Figure 5.6 Actual vs forecasting for KOSPI from 2015 to 2020. In this article, the Korea Composite Stock Price Index (KOSPI), which tracks the prices of all common stocks listed on the Korea Exchange's Stock Market Division (formerly the Korea Stock Exchange), is investigated over the period from 2015 through 2020. Like the S and P 500 in the US, it is the benchmark stock market index for South Korea. In this research, the KOSPI dataset from 2015 to 2020 is investigated. The comparison tables above show the above average of all the methods from the year 2015 to 2020 for KOSPI, where (a) resembles 2015, (b) resembles 2016, (c) resembles 2017, (d) resembles 2018, (e) resembles 2019, and (f) resembles 2020.

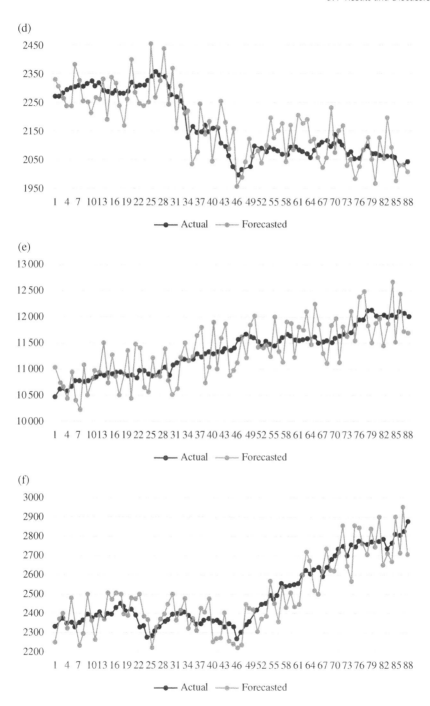

Figure 5.6 (Continued)

long term. This paper assesses the execution of the proposed strategy utilizing RMSE, which is characterized as taking after-

$$RMSE = \sqrt{\frac{\sum_{i=1}^{n}\left(Actualvalue_i - Forecastedvalue_i\right)^2}{n}}$$

where n signifies the number of days required to be forecasted. Table 5.1 shows the comparison of the RMSE and the normal RMSE for diverse strategies and also shows the RMSE values of TAIEX 2015–2020 in Figure 5.4 utilizing genuine information and forecasted information. Figure 5.5 shows the RMSE values of BSE 2015–2020 utilizing genuine information and forecasted information, and Figure 5.6 shows the RMSE values of KOSPI 2015–2020 utilizing real information and forecasted information.

5.4.1 Comparison Table for the Method TAIEX

Table 5.2 shows the RMSE comparison for the TAIEX index from 2015 to 2020. Six different earlier methods have been taken for the RMSE comparison to show the effectiveness of the proposed method.

5.4.2 Comparison Table for Method BSE-SENSEX

Table 5.3 shows the RMSE comparison for the BSE-SENSEX index from 2015 to 2020. Six different earlier methods have been taken for the RMSE comparison to show the effectiveness of the proposed method.

5.4.3 Comparison Table for Method KOSPI

Table 5.4 shows the RMSE comparison for the KOSPI index from 2015 to 2020. Six different earlier methods have been taken for the RMSE comparison to show the effectiveness of the proposed method.

Table 5.1 RMSES and average RMSE comparison for various indices.

Year/method	2015	2016	2017	2018	2019	2020	Average
TAIEX	1121.78	1060.93	586.07	1061.89	1049.69	1999.96	1146.72
BSE	2614.9	4071.01	4197.88	2968.98	3124.15	6202.34	3863.21
KOSPI	93.71	280.09	135	98.07	154.33	397.76	193.14

Table 5.2 RMSE comparison for the proposed method with the earlier methods for TAIEX.

Year/method	2015	2016	2017	2018	2019	2020	Average RMSE
Huarng (2007)	1110.34	1040.78	576.02	1032.7	1039.78	1890.9	1115.08
Chen (1996) (fuzzy time series model)	1118.12	1030.12	560.15	1028.23	1020.21	1809.12	1094.32
Yu and Huarng (2008) (univariate conventional regression model)	1106.38	1003.18	549.5	1020.85	1021.42	1887.71	1098.07
Yu and Huarng (2008) (univariate neural network model)	1126.3	1013.18	540.9	1010.85	1035.42	1855.71	1097.06
Yu and Huarng (2008) (bivariate conventional regression model)	1108.12	1006.17	568.13	1025.23	1030.21	1867.12	1100.83
Yu and Huarng (2008) (bivariate neural network model)	1117.12	1034.12	566.15	1023.23	1039.21	1859.12	1106.49
Proposed method	1112.21	1031.21	556.13	1012.32	1040.31	1867.2	1103.23

5.5 Conclusion

This study suggests a brand-new gate control unit based on research on the RNN and LSTM. The RNN method with LSTM demonstrates an algorithmic approach to analyzing time arrangement information and treating the basic process. In this way, the RNN with LSTM was able to make more precise expectations on stock cost movements compared to the CART demonstrate. This is often since the LSTM model, by its nature, uses a profound learning approach and is good at handling successive information, and extracts useful information while dropping pointless data and it comes with a superior result with benefit. Not like past strategies, this approach is all about employing a factual strategy to look at the determining execution of both the proposed strategy and other strategies. To begin with the case of TAIEX estimation, the proposed strategy is found to offer superior determining execution than the other strategies displayed sometime recently in 2017 and

Table 5.3 RMSE comparison for the proposed method with the earlier methods for BSE-SENSEX.

Year/method	2015	2016	2017	2018	2019	2020	Average RMSE
Huarng (2007)	2613.6	4060.91	4187.8	1050.78	1039.4	1900.78	2475.54
Chen (1996) (fuzzy time series model)	2608.12	4050.12	4160.15	1028.23	1032.21	1889.12	2461.32
Yu and Huarng (2008) (univariate conventional regression model)	2606.38	4037.18	4149.5	1040.85	1020.42	1890.71	2457.5
Yu and Huarng (2008) (univariate neural network model)	2611.3	4043.18	4140.9	1030.85	1037.42	1885.71	2458.22
Yu and Huarng (2008) (bivariate conventional regression model)	2610.12	4036.17	4168.13	1005.23	1030.21	1867.12	2452.83
Yu and Huarng (2008) (bivariate neural network model)	2607.12	4034.12	4166.15	1013.23	1019.21	1859.12	2449.82
Proposed method	2612.45	4057.45	4149.1	1006.1	1021.01	1898.03	2457.35

Table 5.4 RMSE comparison for the proposed method with the earlier methods for KOSPI.

Year/method	2015	2016	2017	2018	2019	2020	Average RMSE
Huarng (2007)	90.67	278.04	130	95.05	145.29	380.73	186.63
Chen (1996) (fuzzy time series model)	88.12	190.12	120.15	68.23	140.21	289.12	149.32
Yu and Huarng (2008) (univariate conventional regression model)	76.38	183.18	119.5	50.85	129.42	297.71	142.84
Yu and Huarng (2008) (univariate neural network model)	86.3	243.18	110.9	90.85	141.42	255.71	154.72
Yu and Huarng (2008) (bivariate conventional regression model)	58.12	266.17	108.13	85.23	130.21	267.12	152.49
Yu and Huarng (2008) (bivariate neural network model)	67.12	234.12	126.15	63.23	139.21	259.12	148.15
Proposed method	85.6	221.09	121.42	89.04	139.31	290.03	157.74

shows no contrast after 2017. Besides, no factual comparison is found between the strategies created after 2017 and the proposed strategy.

One of the reasons is that the proposed strategy as it was used to estimate the TAIEX, whereas other strategies utilize one or more variables, which may progress the determining execution. Although the proposed strategy as it we reemployments one figure, a conventional procedure for dividing the approach, and second-order record the information to estimate the stock file and from the estimating comes about of this paper utilizing three databases of TAIEX, BSE, and KOSPI, the proposed strategy is found to offer superior estimating execution. Currently, to construct the numerous straight models of the dataset, a novel high-order time arrangement of these model calculations is utilized to gather information by their direct connections rather than shapes, which builds a more appropriate direct demonstration than other group-based models. The proposed model calculation is compared with that of the other three model calculations and built the fines treasonable direct demonstrate which gives the accurate result. From a workable point of view, for speculators interested in choosing between measurable strategies, or machine learning techniques, or profound learning procedures for determining which stocks to purchase, we propose the speculators to utilize the LSTM model for their stock cost forecasts because of the deep learning approach of the LSTM shown.

At last, this paper plans a determining show based on the ANN, which calculates the weight of related numerous straight models, as well as its learning calculation. The proposed estimation is compared with another FTS-based demonstration on the TAIEX, BSE, KOSPI, and it gave distant better; a much better; a higher; a stronger; an improved; and a much better determining precision than others. Also, by comparing with estimating models that are FTS-based, the outcomes also propose that the proposed determining demonstration seems to handle the inadequate, uncertain information, and distant better; a much better; a higher; a stronger; and an improved stronger robustness than this. The RMSE estimation of TAIEX is 1146.72, BSE is 3863.21 and KOSPI is 193.14 from table 5.1. It is proposed that advance research be performed to better understand the execution of the LSTM demonstrated in other places of market.

Acknowledgement

I need to thank Shri Partha Pratim Deb for his assist for the duration of this work. I moreover want to thank the analysts for introducing positive grievance and ideas. Their knowledge and grievance significantly display the contemplations communicated on this paper.

References

Chen, S.M. (1996). Forecasting enrollments based on fuzzy time series. *Fuzzy Sets and Systems* 81 (3): 311–319.

Cho, K., Van Merriënboer, B., Gulcehre, C. et al. (2014). Learning phrase representations using RNN encoder-decoder for statistical machine translation. *arXiv Preprint arXiv* 1406: 1078.

Chung, J., Gulcehre, C., Cho, K., and Bengio, Y. (2014). Empirical evaluation of gated recurrent neural networks on sequence modeling. *arXiv Preprint arXiv* 1412: 3555.

Gers, F.A., Schmidhuber, J., and Cummins, F. (2000). Learning to forget: Continual prediction with LSTM. *Neural Computation* 12 (10): 2451–2471.

Gers, F.A., Eck, D., and Schmidhuber, J. (2001). Applying LSTM to time series predictable through time-window approaches. *Artificial Neural Networks – ICANN 2001: International Conference Vienna, Austria, Proceedings 11*, Berlin Heidelberg (21–25 August 2001), pp. 669–676. Springer.

Hochreiter, S. and Schmidhuber, J. (1997). Long short-term memory. *Neural Computation* 9 (8): 1735–1780.

Hopfield, J.J. (1982). Neural networks and physical systems with emergent collective computational abilities. *Proceedings of the National Academy of Sciences United States of America* 79 (8): 2554–2558.

Huarng, K.H., Yu, T.H.K., and Hsu, Y.W. (2007). A multivariate heuristic model for fuzzy time-series forecasting. *IEEE Transactions on Systems, Man, and Cybernetics, Part B (Cybernetics)* 37 (4): 836–846.

Malkiel, B.G. (1989). Efficient market hypothesis. In: *Finance*, London: Palgrave Macmillan UK, 127–134.

Ni, R. and Cao, H. (2020). Sentiment analysis based on GloVe and LSTM-GRU. *2020 39th Chinese control conference (CCC)*, Shenyang (27–29 July 2020), pp. 7492–7497. IEEE.

Wei, X., Zhang, L., Yang, H.Q. et al. (2021). Machine learning for pore-water pressure time-series prediction: application of recurrent neural networks. *Geoscience Frontiers* 12 (1): 453–467.

Yu, T.H.K. and Huarng, K.H. (2008). A bivariate fuzzy time series model to forecast the TAIEX. *Expert Systems with Applications* 34 (4): 2945–2952.

Zheng, J. and Zheng, L. (2019). A dictionary-based convolutional recurrent neural network model for sentiment analysis. *CISCE*, Haikou (5–7 July 2019). IEEE.

6

Study and Analysis of Machine Learning Models for Detection of Phishing URLs

Shreyas Desai[1], Sahil Salunkhe[2], Rashmi Deshmukh[3], and Shital Gaikwad[4]

[1,2,3] *Department of Technology, Shivaji University, Kolhapur, Maharashtra, India*
[4] *Department of Computer Science, Shivaji University, Kolhapur, Maharashtra, India*

6.1 Introduction

With digitalization, almost all the services are at our doorstep, most of them being mailboxes, e-commerce, banking, social media, food services etc. A majority of people have started using these services at an increasing pace. These services ask you to authorize yourself in order to use them; this involves sharing your personal information to them. As for the legitimate services, this is not needed, but for the forged ones, this may even lead to a critical situation. This is when the phishing begins; it has been for the last three decades, and many users have got a problem with this.

Phishing is a type of social engineering attack which is frequently used to get the customer's credentials. Fraudsters trick customers into typing credentials, credit card, or maybe bank account numbers on those fraudulent phishing sites which they host on some hosting service providers. As at least one user gets tricked and clicks on such websites, the DNS server also registers such fraudulent websites on their database. As it comes beneath the social engineering act, it varies from user to user on how they get lured in. The fundamental vector a fraudster can attempt is to first speak about the similar interest the user has followed by a phishing activity like mail or website. The phishing attacks are so easy to drag out relying upon the user and have one of the worst results feasible. One may become completely bankrupt. Not every time it is about stealing credentials; some may even ask to click on a malicious link which can implant a

Machine Learning Applications: From Computer Vision to Robotics, First Edition.
Edited by Indranath Chatterjee and Sheetal Zalte.
© 2024 The Institute of Electrical and Electronics Engineers, Inc.
Published 2024 by John Wiley & Sons, Inc.

backdoor in a company's system which till detected allows the fraudster to get the whole insider. The possibilities are endless and need to be stopped.

Phishing can be defined as luring the users into divulging their personal information to the forged services acting like legitimate ones. It is considered a cybercrime, and there are various types of phishing attacks out there like Vishing, Whaling, Spear Phishing, and Email Phishing. This paper specifically focuses on website phishing and implements the idea of phishing URL detection using ML.

Phishing is derived from the word "phish" which means a fraud and is a serious issue of data security. There are multiple technologies developed over the period of time to stop phishing attacks like the blacklist method which maintains such a database of phishing URLs and blacklists them on the DNS. The heuristic-based method uses feature extraction to find the phishing URLs, but it is also the easiest one to bypass once an attacker knows the features used. Another being the content-based anti-phishing which uses visual similarities to identify the contents of phishing websites from legitimate websites by analyzing the similarity of the contents.

There have been a few examinations attempting to take care of phishing issues. In this paper, we have implemented different machine learning models for detection of phishing URLs based on the features of the URL such as length, paths in URL, and subdomains present. We have used a custom dataset which is created using URLs from the PhishTank dataset for phishing URLs and the Alexa dataset for legitimate URLs. At the end, we have compared the performance of different models on the dataset based on performance metrics like accuracy and precision.

6.2   Literature Review

Hannousse and Yahiouche (2021) proposed a novel method which finds features from the URL. The author measures the performance by using a dataset containing active phishing attacks and compares it further with Google Safe Browsing (GSB). We can also see the four-model (SVM, RF, LR, and KNN) approach which uses only nine features based on the lexical properties of URLs and produces an accuracy of 99.57% [3]. Similarly, Khan and Rana (2021) anticipated a 10-feature approach, and the models were used with accuracy LSTM 98.67%, DNN 96.33%, and CNN 97.23%. After analyzing few more papers, we found out that even DT is best suited to find out such URLs Pujara and Chodhari (2018) compared four models LR, DT, RF, and SVM on the basis of six factors ("absolute error," RMSE, "sensitivity," "training dataset," "time," "cohen_kappa_score" and "roc_auc_score") among 16 factors.

Some of the papers Mahajan and Siddvatam (2018) implemented a hybrid solution of multiple approaches (heuristic features, visual features, and various approaches feeding these distinct features to machines). Bu and Cho (2021) used a hybrid of neural networks and logic programming which has the best prediction level even for zero-day phishing by using the CNN–LSTM-based triplet network.

The KNN too seems to be effective against such kinds of URLs as Assegie (2021) proposed the KNN model with an accuracy of 85.06% using 106 observations for testing the model on phishing detection. Using accuracy metrics for experimental results shows that the model is effective against phishing attacks. Sahingoz et al. (2019) have implemented seven different algorithms (DT, Adaboost, K-star, kNN ($n = 3$), RF, SMO, and Naive Bayes) to improve the accuracy.

6.3   Methodology

A great deal of the works that we have reviewed as of recently uses URL-based features along with the HTML content inside the website to correctly distinguish whether a website is phishing or not. The proposed work has been depicted in the figure below.

6.3.1   Proposed Work

As shown in Figure 6.1, we have started with the phishing URL dataset from the PhishTank.org website. After the dataset is downloaded, we have checked the shape of the dataset (features present and number of rows). We have reduced the number of elements in the feature vector from the already given dataset to the ones which we require. The dataset of top one million websites is collected from the Kaggle data servers. With this, domain names of the top 1million websites are obtained.

After this a python script is executed to scroll the top 550 website domains present in the list to find redirecting URLs present on each of the domains. A list is obtained after scrolling through all these website domains containing all the redirect URLs which can then be treated as a set of benign/legitimate URLs. Lastly, after obtaining two different CSV files, one containing phishing URLs and another containing Legitimate URLs, both the CSVs are combined and shuffled. The model training part is done on this "combined.csv" file.

6.3.2   Traditional Methods

Regardless of whether the site is phishing or legitimate is decided based on the URL features of the website. Different features from the URL are extracted based on its structure. Various features can be used for detecting phishing of

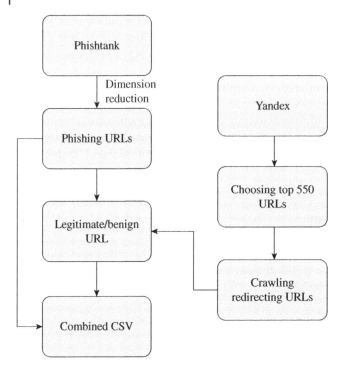

Figure 6.1 Proposed system architecture.

URLs. The traditional methods for phishing detection were Blacklist Method, Heuristic-based model, and visual similarity-based methods. These techniques are as follows-

6.3.2.1 Blacklist Method
Blacklist method is a traditional method which uses a pre-maintained database consisting of phishing URLs which are identified and updated regularly. In this method, the browser simply checks if the URL is already present in the database or not. The drawback of this method is that it cannot keep up with the continuous creation of the new phishing URLs as keeping the database updated every second is not possible. Also, it cannot automatically identify if the URL is phishing or not.

6.3.2.2 Heuristic-Based Model
Heuristic models are modified blacklist methods based on the pre-fed algorithm where the system is able to automatically identify if a URL is phishing or not. But this system is easy to bypass once the attacker knows the algorithm which is used to detect phishing URL.

6.3.2.3 Visual Similarity

Commonly, a phisher attempts to trick a user by using visual similarities from legitimate websites and causing the user to accept that he is visiting a genuine site. In the visual similarity method, the algorithm attempts to analyze the visuals from the phishing websites with those of legitimate websites. The downside of this framework is that it takes more time to compare images and regardless of whether there is a little change in the pictures, the algorithm may fail.

6.3.2.4 Machine Learning–Based Approach

Aside from the conventional methodology stated in the above methods, we can use machine learning approaches to deal with identifying a phishing URL. The different features we can use while detecting using machine learning are HTML content-based features, NLP-based detection, and URL-based detection. In our project, we have attempted to minimize the overload on the system by minimizing the features used for faster training and trying to achieve as much high accuracy as possible.

6.4   Results and Experimentation

The implementation of the project is done with a system having Processor – i5 9th generation, with 8GB RAM and Secondary Memory – 1TB HDD. The operating system is Windows 11, and experimentation is carried out with IDE – Jupyter Notebooks.

6.4.1   Dataset Creation

For the implementation; we have created a custom dataset. This dataset is created with the help of PhishTank and Alexa. PhishTank is used for the collection of phishing URLs, whereas Alexa is used for the collection of legitimate URLs. As mentioned in the proposed work, for the creation of the dataset, we have downloaded the phishing URL dataset from the PhishTank (~4200 URLs). Pre-processing is carried out by eliminating unwanted columns from the dataset.

The Alexa dataset contains one million URLs. As the Alexa dataset records the top sites all over the world, most of the websites incorporate names like "googl.com," "amazon.com," and "facebook.com." These website URLs are not useful for train-ing our data because we cannot extract any useful features from the previously mentioned websites. To avoid this, we have executed a python script which visits the first 550 websites and scrolls their webpages to discover the redirecting URLs present on the webpage. With this we have got a large set of URLs (~17700) with multiple features and which can be used to train our model.

6.4.2 Feature Extraction

We have used URL-based features for training our model so that the overhead on the model is reduced. There are a number of features which can be used based on URL from which it can be deduced whether it is phishing or not. From the features available, we have used the features shown in Table 6.1.

6.4.3 Training Data and Comparison

For the implementation, we have used six different machine learning models. Comparison of the learning models is based on the accuracy score, precision, recall, and F-1 score on the training as well as testing data.

6.4.3.1 XGB (eXtreme Gradient Boosting)
XGBoost is an algorithm which has emerged recently, dominating the applied machine learning for structural or tabular data. XGBoost is an extended version of gradient boosting decision trees which improves speed and performance.

6.4.3.2 Logistic Regression (LR)
LR is a supervised ML algorithm used to forecast definite dependent variables based on independent variables. Logistic regression works best when there are two discrete values i.e. True or False, 0 or 1, and Yes or No. It is used for classification problems.

Table 6.1 Features table.

No.	Feature	Values in the dataset
1	Length of URL	Integer
2	Whether IP address is present or not	0 and 1
3	Whether any shortening service is used or not	0 and 1
4	Special character counts in URL	Integer
5	Presence of "//" in URL other than "[http/https:]//" in URL	0 and 1
6	Phish hints in the URL	Integer
7	Common term count in the URL	Integer
8	Digit ratio	Float
9	URL depth	Integer
10	Abnormal subdomain	0 and 1

6.4.3.3 RFC (Random Forest Classifier)

RFC is a collection of many decision trees. These decision trees are formed for different small subsets of the dataset, and the average of the trees is taken to improve the predictive accuracy. This model gives more accurate predictions than a single decision tree. The higher the number of trees in RFC, the higher the accuracy and lesser overfitting.

6.4.3.4 Decision Tree

The decision tree classifier is a supervised learning algorithm which is best suited for classification problems. In decision trees, there are leaf nodes and decision nodes. Decision nodes are nodes which make decisions and branches into leaf nodes. Leaf nodes do not have branches, and they represent the results of the tree.

6.4.3.5 SVM (Support Vector Machines)

SVM is a supervised learning algorithm which can be used for regression as well as classification problems. SVM finds a hyper plane in an N-dimensional space for classification of different data values. The dimensions of the hyper plane depend upon the number of input features.

6.4.3.6 KNN (K-Nearest Neighbors)

KNN is a widely used regression and classification supervised learning algorithm. The working of KNN is simpler to understand. In KNN, we simply pick a data point and put it in the graph with a feature matrix as the axis. We then simply set the number of K (no. of neighbors) around the data point. What KNN does is it classifies the data point based on the similarities of the data points which are already present. Then, it checks the count of the data points in each category and assigns the new data point to the category with the maximum number of neighbors.

6.5 Model-Metric Analysis

While doing an analysis of different machine learning models, we take into consideration various factors such as its accuracy, time taken for the model to execute, and how much hyper-parameter tuning can be done to make the model work at its best. We have taken some of these factors to check the performance analysis of the models. We will be doing the analysis of the models based on the confusion matrix for different data parts in the dataset.

Table 6.2 shows a comparison of different models based on the above-mentioned performance metrics on both training as well as testing data.

From Table 6.2, we come to know that except the logistic regression model, most of the models work efficiently on the training data. But while using a

Table 6.2 Comparison of models based on different performance metrics.

	Accuracy		Precision		Recall		F1 Score	
Model	**Train**	**Test**	**Train**	**Test**	**Train**	**Test**	**Train**	**Test**
XGBoost	94.98	88.02	92.15	83.13	95.35	86.58	93.73	84.82
LR	76.06	75.94	67.08	66.32	72.14	71.15	65.92	68.92
RFC	94.81	88.32	91.88	83.13	95.20	87.26	93.51	85.14
DT	95.00	86.91	91.17	80.99	96.36	85.69	93.69	83.27
SVM	91.10	85.53	86.70	79.72	91.01	83.58	88.80	81.60
KNN	93.28	83.74	92.28	77.75	91.30	81.80	91.79	79.38

machine learning model in a real-time scenario, we have to take into consideration how well the model can perform on unseen data. Hence based on the above table, we will focus more on the columns having metric analysis for test data. Figure 6.2 shows that the XGBooster model and the Random Forest Classifier have the best test data accuracy among the given machine learning models. The precision and recall of these models are also noteworthy as based on the precision and recall, we come to know how much of the predicted values are false negative or false positive, as these values have a greater effect on how well our model works, as shown in Figures 6.3–6.5.

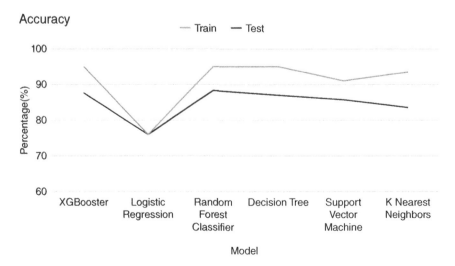

Figure 6.2 Training and test data accuracy.

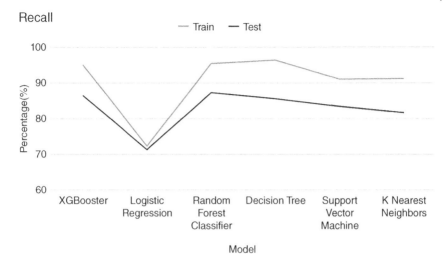

Figure 6.3 Training and test data recall.

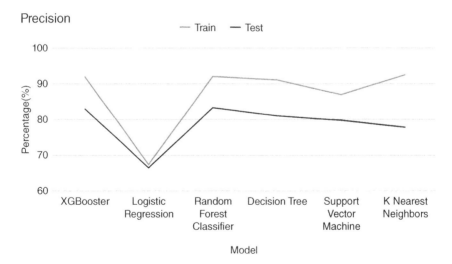

Figure 6.4 Training and test data precision.

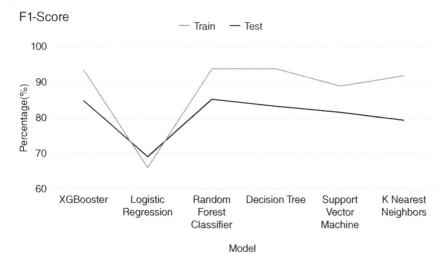

Figure 6.5 Training and test data F1score.

6.6 Conclusion

In this paper, we began with an outline about phishing attacks present in real-time scenarios and traditional methods present to detect and invalidate them. We then studied different approaches used for detection of the phishing URLs. We have tried to overcome some of the drawbacks of the existing system. For the implementation, we have created a custom dataset, and then some features are extracted from it. Then, different machine learning models are applied to the dataset. Finally, we have compared these models with various performance metrics like accuracy of the algorithm, precision score, recall value, and F1-Score. Experimentation shows that comparatively, the Random Forest Classifier and XGBoost model perform better to identify a phishing URL with the use of a minimum number of features.

References

Assegie, T.A. (2021). K-nearest neighbor based URL identification model for phishing attack detection. *Indian Journal of Artificial Intelligence and Neural Networking* 1 (2): https://doi.org/10.35940/ijainn.B1019.041221.

Bu, S. and Cho, S. (2021). Integrating deep learning with first-order logic programmed constraints for zero-day phishing attack detection, Toronto (6–11 June 2021), pp. 2685–2689. https://doi.org/10.1109/ICASSP39728.2021.9414850.IEEE.

Hannousse, A. and Yahiouche, S. (2021). Towards benchmark datasets for machine learning based website phishing detection: an experimental study. *Engineering Applications of Artificial Intelligence* 104. https://doi.org/10.1016/j.engappai.2021.104347.

Khan, F. and Rana, B.L. (2021). Detection of phishing websites using deep learning techniques. *Turkish Journal of Computer and Mathematics Education* 12 (10): 3880–3892. https://doi.org/10.17762/turcomat.v12i10.5094.

Mahajan, R. and Siddvatam, I. (2018). Phishing website detection using machine learning algorithms. *International Journal of Computer Applications* 181: 45–47. https://doi.org/10.5120/ijca2018918026.

Pujara, P. and Chodhari, M. (2018). Phishing Website Detection using Machine Learning: A Review. *International Journal of Scientific Research in Computer Science, Engineering and Information Technology* 3 (7): 395–399.

Sahingoz, O., Buber, E., Demir, O., and Diri, B. (2019). Machine learning based phishing detection from URLs. *Expert Systems with Applications* 117: 345–357. https://doi.org/10.1016/j.eswa.2018.09.029.

7

Real-World Applications of BC Technology in Internet of Things

Pardeep Singh, Ajay Kumar, and Mayank Chopra

Department of Computer Science and Informatics, Central University of Himachal Pradesh, Dharamshala, Himachal Pradesh, India

7.1 Introduction

Blockchain (BC) technology is designed to support cryptocurrency and enable transactions without a mediator. Bitcoin is the first cryptocurrency based on a BC (Vivekanadam 2020). BC technology works on distributed ledger technology (DLT) constructed from blocks, evolving lists of records. An individual system inside the BC network is called a node. Three categories are used to categorize it: publishing node, complete node, and lightweight node. It publishes new structures in the publishing node. The complete node holds the entire BC and verifies that all transactions are legitimate. A lightweight node does not keep copies of the BC; transactions are sent to full nodes. Peer-to-peer networking, game theory, and cryptography are all combined in DLT (Biswas and Muthukkumarasamy 2016).

The Internet of Things (IoT) is the linking of intelligent objects to gather information and make informed choices. Because of the absence of inherent security safeguards, the IoT exposes safety and privacy risks. Because of the "security by design" feature, BC can effectively help solve critical safety needs in the IoT. "Most of the IoT's architectural flaws can be fixed with the help of BC characteristics, including consistency, visibility, audibility, file encryption, and operational resilience (Panarello et al. 2018)." The IoT is a network of electrical appliances, mechanical and digital machinery, items, and people with unique identifiers and the capability to transmit data across human-to-human or desktop relationships. Another feature of the Internet is that IoT objects, like people and computers, have IP addresses (identifiers) that can be assigned and used to send data to arbitrary objects or networks (Konashevych 2020). Since Kevin Ashton coined the phrase "Internet of Things" in 1999, it has developed from a straightforward idea into one of the most potent forces behind commercial development. IoT integration with big data, cloud computing, and machine learning has made it the cornerstone on which online information services are constructed. IoT gadgets today include anything from wearable technology to platforms for developing hardware (Doguet 2013; Hellani et al. 2019). "In 2019, the frequency of IoT-connected gadgets surpassed 5 billion, which is expected to keep increasing until it reaches 29 billion by 2022 (Mistry et al. 2020)." According to the National Intelligence Council and McKinsey Global Institute, by 2025, commonplace items, including furniture, food packaging, and papers, will act as Internet nodes. The term "blockchain" has become the official label for the database tracking cryptocurrencies. The BC's resilient architecture protects the distributed ledger, which is its main benefit. The reduced effort, complication, and cost-effectiveness are other advantages of BC technology. BC is a distributed repository of transaction records assessed and maintained by a worldwide computer network. The data are handled by a vast community rather than a single centralized administration, like a bank; no one has authority over it. Nobody can go back in time and edit or wipe a piece of information from any transaction data. The data cannot be changed like a conventional, centrally located database because of the distributed characteristic of the BC's structure and participant assurances. While a typical centralized database is kept on a single server, the BC is spread across all system users. Everyone in the ecosystem has an insight into each other's BC transactions (Sarmah 2018). This chapter delves into the practical applications of BC technology in IoT.

7.1.1 Relevance and Benefits of Blockchain Technology Applications

As stated by PwC's global BC survey 2018, 84% of firms worldwide actively use BC-based technology in some capacity, with 15% successfully implementing it.

PwC polled 600 business people from 15 different countries about their use of BC and their opinions of the technology's potential to determine the status of BC in 2018. According to Deloitte's 2021 Global Blockchain Survey, banks should embrace their inevitable transition to digital technology. Financial leaders suddenly see more and more value in virtual currencies as the future. Figure 7.1 elucidates the various types of BC networks, encompassing both private and public paradigms. It provides a comprehensive distinction between permissioned and permissionless configurations within each type. The figure expounds on the unique attributes of private and consortium BCs as permissioned, emphasizing controlled access and validation mechanisms. Conversely, it highlights the permissionless nature of public and hybrid BCs, emphasizing their decentralized and open characteristics. The following vital advantages BC provides cause it to become popular in the sector.

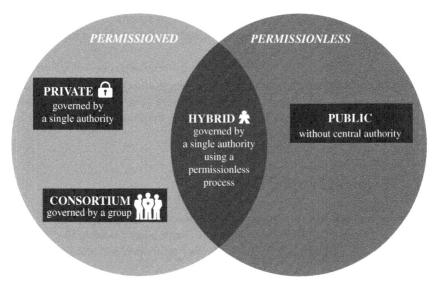

Figure 7.1 Types of Blockchain.

Transparency: Customers are more informed than ever and want more details about the goods they buy. Transparency of transaction history has never been greater than with BC technology. Each participant shares the distributed ledger via the network. Anybody can easily navigate the data on a BC ledger. Everyone on the network can recognize the change and the amended record if a transaction history changes. As a result, everyone has direct exposure to all currency exchange information (Chod et al. 2020).

Traceability: A traceable system enables tracking of the products by providing details about the products (such as originality, components, or locations) during manufacture and distribution. The financial transactions of items are monitored with BC, providing an audit trail to determine the origin of a particular asset. You also gain knowledge about every stop the item makes along the direction. This degree of monitoring can help clients confirm the product's legitimacy and prevent fraud. (Lu and Xu 2017).

Security: By all measures, BC is more secure than any other record-maintenance method. Only a BC network's consensus can update or modify the shared records of transactions. The BC network is costly to hack regarding time. A record is only updated if most nodes, or all of them, concur. When a transaction is accepted, it is encrypted and linked to the previous transaction. Therefore, a single party or an individual cannot change a record. Due to the decentralized characteristic of the BC, no one has the independent authority to update records. Since the technology has been built so that no block can be changed without affecting the rest of the block's data, it becomes useless to try to hack a BC (Li et al. 2020b).

Auditability: Because each transaction is documented within the BC for the time frame of its existence, you may access the audit trail to verify the legitimacy of your asset.

Inexpensive: Because a BC does away with intermediaries and third parties, firms can save money. You can decide on the terms and conditions of exchange without anyone else's help. Enabling everyone else to access a single, unchangeable rendition of the ledger minimizes the time and money spent on documentary evidence and its modifications.

7.2 Review of Existing Study

Incorporating the insights gained from our comprehensive review of the literature, as presented in Table 7.1, we have synthesized a cohesive overview of the relevant work within our book chapter on Real-World Applications of BC in IoT.

Table 7.1 Review of the existing study.

Title	Author(s)	Contribution	Limitation/future scope
"Blockchain with Internet of Things: Benefits, Challenges, and Future Directions"	Ahmed Alenezi, Madini O. Alassafi, Gary B. Wills (Atlam et al. 2018).	This study describes Blockchain integration with IoT, highlighting the advantages and obstacles of integration.	Smart contracts, regulatory laws, security, IOTA
"Blockchain, Internet of things, and artificial intelligence"	N Chilamkurti, T Poongodi, B Balusamy (Chilamkurti et al. 2021).	According to this study, using blockchain will eliminate security concerns in the IoT.	Open issues on blockchain integration with IoT has been discussed for future research directions.
"Blockchain beyond Bitcoin: Blockchain Technology Challenges and Real-World Applications"	M Memon, SS Hussain (Memon et al. 2018).	The solution provided in this study allows for implementation on both small-scale and larger network levels.	The major issue that needs to be solved is the secure deployment of blockchain-based technology on small-scale applications.

Table 7.1 (Continued)

Title	Author(s)	Contribution	Limitation/future scope
"Blockchain Platform for Industrial Internet of Things"	Arshdeep Bahga, Vijay K. Madisetti (Bahga and Madisetti 2016).	This study described a test case of the BPIIoT platform based on the Beaglebone Black SBC, a connectivity board utilizing the Ethereum Blockchain network and the Arduino Uno.	The BPIIoT platform can be implemented and demonstrated for more practical solutions such as device self-service and on-demand production.
"Blockchain in Internet of Things: Challenges and Solutions"	Ali Dorri, Salil S. Kanhere, and Raja Jurdak (Dorri et al. 2016).	They presented that blockchain-based IoT architecture has consistent overhead at worst and best; most of its transactions are proportional to the number of cluster centers in the network instead of the number of nodes.	Questions remain unanswered about lowering the threat of denial-of-service assaults, modification threats, and 51% damage inflicted by establishing distributed trust.
"Evolution of Internet of Things from Blockchain to IOTA: A Survey"	Mays Alshaikhli, Tarek Elfouly, Omar Elharrouss, Amr Mohamed and Najmat Hottakath (Alshaikhli et al. 2021).	IOTA is presented as the new technology for IoT that overcomes the shortcomings of blockchain for application in IoT.	Many open research issues must be addressed to ensure that IOTA is a good fit for IoT devices in terms of security and effectiveness.

7.3 Background of Blockchain

7.3.1 Blockchain Stakeholders

It is critical for substantial IoT applications to share credible data across credible stakeholders. Inherently, organizations and relevant stakeholders in IoT have a dearth of effective partnerships that pose significant challenges to the preceding eyesight, such as ensuring that data in the physical realm can be accurately introduced into the IoT information world, ensuring the trustworthiness of the organizations' identities in IoT, ensuring the reliability of data and the consistency of identity, and ensuring the secure transmission of messages in the absence of a reliable party (Shi et al. 2021).

7.3.2 What is Bitcoin?

The first cryptocurrency built on a BC is called Bitcoin. A Bitcoin is a digital asset with no literal depiction, legal currency status, government, or other legal organization backing it and no central bank controlling its production. Although Bitcoin transactions are private, no traditional financial organizations are involved. The Bitcoin ecosystem is entirely decentralized, with users handling all aspects of transactions, unlike past digital currencies with some central controlling individual or organization. Bitcoin is a peer-to-peer ledger framework for contracts, sales, agreements, and transaction recording. A group of transactions is what is meant by a ledger. The real benefit of BC is that it is hard to launch an assault against the network because doing so would require compromising 51% of its systems (Hou 2017). In January 2009, a computer scientist using the pseudonym Satoshi Nakamoto created Bitcoin. His invention is a fully accessible peer-to-peer digital currency. No traditional financial institutions are participating in transactions in the private Bitcoin network.

7.3.3 Emergence of Bitcoin

Users of Bitcoin are allegedly offered three advantages: reduced transaction fees, enhanced privacy, and a lack of inflation-related loss of purchasing power. On the Bitcoin BC, a digital address is attributed with information representing electronic currency. Bitcoin users can verify the information and transfer ownership to another user digitally. The Bitcoin BC documents this transaction openly, allowing each network user to independently confirm or refute the authenticity of a transaction. A distributed group of people stores and handles the Bitcoin BC collaboratively. When combined with specific cryptographic techniques, the BC is resistant to later attempts to alter the ledger. (Sharma et al. 2019). About seven billion Internet-connected gadgets were used worldwide in 2018 (Boncea et al. 2019). They highlight that the future will produce by incorporating technologies that engage with the surrounding environment of people (Shammar and Zahary 2019). The estimate is that there will be 35 billion by 2025 (Zhidanov et al. 2019). Others estimate that 50 billion IoT devices could exist by 2025.

7.3.4 Working of Bitcoin

Bitcoin users can benefit from lower transaction fees, more anonymity, and long-term preservation against asset prices and purchasing power loss. However, it has several drawbacks that might prevent its use from spreading. These include the significant price volatility of Bitcoins, the unpredictability of security against fraud and theft, and a long-term de-escalation bias that promotes Bitcoin hoarding. To validate transactions and control the creation of the money itself, Bitcoin is based on cryptographic concepts. Every transaction is documented on a decentralized distributed ledger visible to the network's computers but does not divulge sensitive information about the participants. Each Bitcoin, as well as each user, has a unique encrypted identity.

7.3.5   Risk in Bitcoin

Bitcoin is right now the most popular virtual currency, but there will always be some challenges, as with any new dimension. When investing in Bitcoin, there are a lot of severe risks.

The market is erratic and unstable: The Bitcoin price is rapidly fluctuating. A small investment made over the long term is advised to prevent a significant loss.

Fraud: Fraud in Bitcoin trading is a potential problem besides hacking. Some bitcoin exchanges can be scams; the risk results from a lack of security for large investors. Although there are procedures to combat online fraud, security is a significant concern.

Cyber-theft: Since Bitcoin relies on technology, numerous stories indicate losses occurred during mining or trading. Hackers target exchanges, but the wallet program itself is secure.

Absence of rules: The Bitcoin market operates with no restrictions. The government does not have a firm position on Bitcoin or other cryptocurrencies. Because Bitcoin trading is tax-free, it may be a desirable investment alternative. Uncertainty exists regarding the near-term future of the Bitcoin market.

Reliance on technology: Bitcoin is created, traded, and mined digitally using smart wallets. There is no tangible collateral to support it. In contrast to Bitcoin, when we buy gold, property, or mutual funds, we acquire an item that serves as tangible proof that can be sold.

Financial setback: Investment in Bitcoin causes a bubble economy. Bitcoins are useless when the bubble pops. There may be many Bitcoin owners who want to sell. There is no profit from the investment, which results in a terrible financial loss.

7.3.6   Legal Issues in Bitcoin

Bitcoin presents various regulatory and legal issues, including its ability to aid in money laundering, how it will be governed by federal securities law, and how it will fit into the foreign currency trading regulation framework. Each nation has established unique regulations for handling both domestic and foreign currencies. Special laws have been passed to regulate foreign currencies. Citizens can conduct business under any rules and regulations that apply to foreign money. The regulations link to other laws in the following ways: business legislation and investment caps. No bank issues or controls Bitcoins. They are created through mining, a computer-generated procedure. Despite having no connection to the government, cryptocurrencies operate as one-to-one payment systems because they lack a physical counterpart. There are currently no consistent international rules governing Bitcoin. (Krishnan et al. 2020).

With Bitcoin's self-managed money supply, a central authority is also unnecessary, which might shield it against volatility and political influence. Although the system undoubtedly offers potential for criminal mischief that cannot be disregarded, the cryptocurrency is still in its infancy to have established that its business model is viable and, as a result, that it will continue to support illegal actions.

Additionally, the Bitcoin network has offered a novel way to exchange value that might be the basis for future breakthroughs. Therefore, any regulatory response should be initially circumscribed and meticulously measured. The solution is not prohibition. Due to the distributed structure of Bitcoin, even if prohibition ends up being the right decision, it would probably never solve the issues for which it was implemented.

However, more crucially, hand-to-hand cash transactions are utilized for illegal activity just as frequently. Still, lawmakers and law enforcement organizations focus on the offenders rather than the exchange method. The BC for Bitcoin keeps track of every transaction made on the network. Law enforcement organizations might be able to use this public ledger in conjunction with "advanced network analysis techniques" to track down any modest criminal actors by employing digital currency for various purposes. (Doguet 2012).

Government policy measures must be aimed at the Bitcoin exchanges regarding substantial criminal operations. Regardless of whether cryptocurrency exchanges are covered by the most recent amendment to money service organizations' governmental rules and standards, there is no longer any question. The specifications were made to thwart the same massive criminal activity noted as a big problem in the Bitcoin system. The wisest course of action moving forward, until the real nature of the Bitcoin economic system is fully realized, is to ensure that these exchanges adhere to the existing regulatory structure, even though it may only have the intended result on a local scale. (Doguet 2013).

7.4 Blockchain Technology in Internet of Things

7.4.1 Need of Integrating Blockchain with IoT

There are already 5 billion connected gadgets in the IoT, and by 2022, there will be 29 billion (Barboutov et al. 2017). On the Internet, every device generates and trades data. Therefore, it is simple to grasp that we are referring to a significant and ongoing data creation, given this enormous number of devices. It must be easier to address the fundamental security concerns for such a sizable information infrastructure. IoT's data integrity and distributed design are critical challenges. Typically, each node in an IoT ecosystem represents a potential bottleneck that could be used to conduct online assaults like distributed denial-of-service (Kolias et al. 2017). The IoT has several important uses, including decision support systems. Timely judgments can be made using the data compiled from the array of sensors. Therefore, it is crucial to defend the system against injection assaults that attempt to insert fake measures and influence decision-making (Gubbi et al. 2013).

Many vulnerabilities in an IoT ecosystem affect data integrity, confidentiality, and privacy. Because of this, ICT sector researchers and developers chose to incorporate "security by design" technologies into a setting that would allow IoT to overcome the constraints. One such system is BC, which automates transaction authorization and automation using smart contracts and provides legitimacy, non-repudiation, and consistency by default. (Panarello et al. 2018). IoT systems must deal with the diversity of IoT, poor interoperability, IoT device resource constraints, and security and privacy flaws. IoT systems can benefit from BC technologies' increased interoperability, improved privacy, and enhanced security. Additionally, BC can improve the scalability and dependability of IoT systems (Reyna et al. 2018). We call this integration "BIoT" (BC with the Internet of Things) for simplicity. In contrast to existing IoT systems, BIoT provides the following potential advantages.

7.4.1.1 IoT Data Traceability and Reliability

Data on the BC may be located and validated anytime, anywhere. In the meantime, tracing every historical transaction kept on a BC is possible. This makes examining and confirming the products' authenticity and quality possible. Furthermore, because BC transactions cannot be changed or falsified, the accuracy of IoT data is ensured by the data integrity of BCs.

7.4.1.2 Superior Interoperability

IoT data can be transformed and stored on BCs to enhance IoT interoperability. Various IoT data types are transformed, analyzed, retrieved, condensed, and deposited in BCs throughout this process. Additionally, since BCs are constructed on the P2P layered ecosystem that enables global Internet access, interoperability also manifests in navigating many disjointed networks smoothly.

7.4.1.3 Increased Security

Because BC transactions that contain IoT data are digitally signed and encrypted with data encryption (e.g., the elliptic curve digital signature technique (Johnson et al. 2001)), IoT data can be safeguarded via BCs. Additionally, by automatically updating the firmware on IoT devices to patch security flaws, coupling the IoT network with BC technologies can enhance the safety of IoT systems. (Christidis and Devetsikiotis 2016).

7.4.1.4 IoT System Autonomous Interactions

BC technologies can enable automatic interaction between IoT components or devices. As an illustration, Zhang and Wen (2015) suggest that Distributed Autonomous Corporations (DACs) automate transfers in which conventional stakeholders, such as governments or businesses, are not engaged in the billing. DACs can function independently and without human communication, thanks to smart contracts, which reduce costs.

BC can also provide a secure channel for intelligent devices to communicate with one another (Panarello et al. 2018). Three open-source implementations of distributed records are Hyperledger, IOTA, and Ethereum. They all have the crucial elements of connecting across blocks in common, being highly secure by using cryptography to hash many transactions in a single block. Furthermore, the degree of significance of the IoT differs in terms of how views, particularly smart contracts, are linked.

7.4.2 Hyperledger

The Hyperledger Initiative (www.hyperledger.org) is a cooperative initiative to develop a distributed ledger platform and code base that is enterprise-grade, open, and flexible. It seeks to advance distributed ledger technology by putting forth and establishing a merged open-source framework that could transform how business is conducted worldwide (Cachin 2016). It is coded in general-purpose languages and supports custom use cases with modular consensus. The Hyperledger system does not run on cryptocurrencies. Fabric is an approach that utilizes a variable enrollment concept that can be combined with industry-standard administration. Fabric proposes a new BC topology to permit such adaptation and improves how BC responds to probabilistic reasoning, resource depletion, and execution attacks. It uses network access, and chain code-based decentralized applications, variable agreement with the current implementation of practical byzantine fault tolerance (PBFT), trust in root-cause experts, as well as unbalanced cryptography and computerized marking highlight Secure Hash Algorithm (SHA3) and Elliptic Curve Digital Signature Algorithm (ECDSA) (Cachin 2016).

Open enrollment is possible with most decentralized records. However, Hyperledger's permission components increase security by preventing Sybil attacks, which occur when a harmful substance creates and selects ill-conceived colleagues' Sybil identities that negatively impact the network. Additionally, its smart contract design includes chain code execution, which may self-execute circumstances like when a resource or asset transfers between colleagues in milliseconds. (Yan et al. 2014; Samaniego and Deters 2016; Smith and Christidis 2016). Using PBFT, Hyperledger avoids the probability-based and costly computational hash mining required for proof of work. However, it also trades off instantaneous computation costs with network consumption. Many concurring peers communicated and integrated the transaction after predictable execution, resulting in a comparable block (Cachin 2016).

7.4.3 Ethereum

In Vitalik Buterin's paper (Walker et al. 2005), Ethereum was created to overcome various drawbacks of Bitcoin's programming language. The most significant features include complete Turing completeness, which means that Ethereum supports any computation, even loops. Furthermore, Ethereum pillars the transaction's state and other enhancements to the BC structure. Initially, the Ethereum BC used a proof-of-work-based (PoW) consensus mechanism, allowing Ethereum system components to consent to the current state of the Ethereum BC, including all data collected on the BC network blocking some sorts of economic attacks. However, in 2022, Ethereum abandoned proof-of-work in favor of proof-of-stake. The BC Ethereum features a Turing-complete programming language. Using the abstract layer, anybody may build ownership laws, transaction setups, and state transition techniques. To do this, smart contracts, a series of encryption rules that only activate when specific criteria are satisfied, are employed (Vujičić et al. 2018). Based on a modified version of the Greedy Heaviest Observed Subtree Protocol (GHOST), the Ethereum network uses consensus (Sompolinsky and Zohar 2015). It was made to address the issue of outdated portions in the system. These outdated blocks can develop if one set of miners in a mining pool looks to have more computational power than the rest, implying that blocks from the previous stream will significantly contribute to the system and cause the centralization problem. The GHOST protocol considers stale blocks for selecting the longest chain. By awarding stale block incentives, with the outdated block receiving 87.5% of the billing and the descendant of that outdated block receiving the remaining 12.5%, the centralization problem is resolved. Thus, miners are paid even if the dominant network rejects their block. A GHOST protocol variant used by Ethereum supports up to seven generations (Buterin 2014).

7.4.4 IOTA

However, some fundamental traits prevent BC technology from being applied to IoT collaboration platforms, like the requirement for constantly increasing storage and low scalability. BC provides numerous advantages, including complete decentralization, data integrity, privacy preservation, and anonymity (Hellani et al. 2018). Furthermore, these infrastructures' scalability and transactional costs are now considered significant barriers to implementation in IoT-based industries. The Tangle protocol, employed in the IOTA platform, was created to address this shortcoming. Tangle was mainly created to overcome scalability difficulties in traditional BC-based platforms (Hellani et al. 2019). A "Tangle" comprises a

Table 7.2 Comparison of distributive ledger–IoT technologies.

Characteristics	Hyperledger	Ethereum	IOTA
Block generation (in seconds)	2 s	15 s	12 s
Transaction time	5e-5–0.1 s	10–15 s	120 s
Computational cost	Less	More	Less
Consensus mechanism	PBFT	PoW	N/A
Network access	Permissioned	Permissionless	Both
Appropriate for IoT	Yes	Yes	Yes
Network type	Private	Public/private	Public

Source: Pustišek and Kos (2018), Rawat et al. (2022).

rigorous mathematical basis known as a Directed Acyclic Graph (DAG). Transactions are admitted into the public ledger by Tangle's validation mechanism after two additional randomly chosen transactions are verified as authentic using a Poisson distribution. The system is therefore scalable and independent of mining or trading expenses. IOTA offers some security, but it is questionable how much (Hellani et al. 2019). Table 7.2 presents a comparative analysis among Ethereum, Hyperledger, and IOTA. This comprehensive examination sheds light on the distinguishing features, strengths, and limitations of each BC platform, allowing for a nuanced evaluation of their respective capabilities.

7.5 Challenges and Concerns in Integrating Blockchain with the IoT

Many professionals have mentioned how BC might secure the IoT, but research and development in this field are ongoing. There are several hurdles to implementing BC with IoT. These are the following:

7.5.1 Blockchain Challenges and Concern

BC is a new technology that is currently dealing with numerous issues, including the following:

7.5.1.1 Scalability
The BC is getting heavier as more transactions are being made. The storage capacity of the Bitcoin BC has increased to more than 100 GB. To be validated, every transaction needs to be saved. Furthermore, the Bitcoin network could only

process roughly seven transactions per second because of the initial data volume constraint and the time required to construct a new block, coming up short of the need to handle thousands of transactions. Meanwhile, because block space is limited, many tiny transactions may be slowed because miners prefer high-fee transactions. Large block sizes, on the other hand, would cause propagation to be delayed and result in BC forks. The term "complexity" refers to the fact that it is difficult to understand (Monrat et al. 2019).

7.5.1.2 Privacy Infringement

Because users only utilize addresses instead of real identities, the BC is considered highly secure. Users could generate many addresses in the event of data leaking. Additionally, because the details of all monetary transactions for every asymmetric key are publicly exposed, it cannot ensure transactional privacy. The emergence of quantum computers has created new obstacles, and a suitably powerful hypothetical quantum computer may easily break the most common public-key methods. Thus, anti-quantum algorithms such as blind signatures, aggregated signatures, and ring signatures should be investigated to increase the security of cryptography algorithms. Because BC and intelligent contracts are based on code, they have always been targets for hackers. BC software and intelligent contracts must undergo more stringent and exhaustive testing to improve code security (Zheng et al. 2019).

7.5.2 Privacy and Security Issues with Internet of Things

IoT technology is bridging a critical communication gap between individuals. It allows for more effective communication. Furthermore, it improves people's lives by enabling intelligent systems, innovative agriculture systems, and other smart systems that people require. As helpful as this technology is, criminals try exploiting it by attacking IoT systems and profiting from innocuous sensitive information and data. Consequently, it is essential to establish methods and strategies for protecting IoT systems. As a result, people's sensitive information is protected (Alferidah and Jhanjhi 2020). The IoT's many interconnections and diversity of devices and technology provide potential cyberphysical security flaws that various cyberattackers can access.

On the one hand, sophisticated types of cyberattacks, like zero-day assaults, have primarily grown with the capability of wreaking havoc on the human environment. For example, in December 2015, a power grid in Ukraine was attacked, causing 225 000 people to lose power (McCarthy et al. 1800). On the other hand, hacking tools have now been largely automated, allowing even inexperienced hackers to carry out catastrophic cyberattacks (Grammatikis et al. 2019). IoT system security and privacy have become a difficulty and an essential component of

IoT systems. The risk degree of privacy and security concerns varies. Specific attacks are more hazardous than others.

Furthermore, attacks vary in their origin. Some attacks are internal, while others are external. Attacks might differ in nature, but their negative impact is the same but varies in severity.

7.6 Blockchain Applications for the Internet of Things (BIoT Applications)

Bitcoin, the first BC application, is a type of digital money that conducts transactions through peer-to-peer networks using a public ledger called the BC. BC technology is used in numerous financial applications besides Bitcoin, including smart contracts and Hyperledger. Therefore, applications for BC technology can be made in a broad range of fields as shown in Figure 7.2 (Tasatanattakool and Techapanupreeda 2018). BC is the missing puzzle that will reveal IoT security and dependability flaws. Because BC is decentralized and self-governing, it is suited for use in various settings, like smart homes and intelligent transportation networks. Furthermore, utilizing innovative contracts may empower self-government and eliminate central authority. Furthermore, BC can establish a safe communication channel for intelligent devices (Mistry et al. 2020).

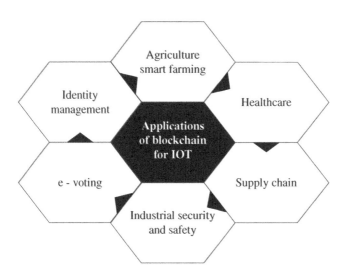

Figure 7.2 Blockchain applications for IOT.

7.6.1 BIoT Applications for Smart Agriculture

Food safety has become a more critical issue around the world. To address food safety challenges from a technological standpoint, consumers require a reliable food monitoring system that can monitor and track the entire life-cycle of agricultural production, including food raw material vegetation, processing, shipping, storing, and selling things. In this work, we propose a reliable, self-organized, open, and environmentally friendly BC-based and IoT-based food monitoring system. It incorporates all participants in an intelligent agriculture ecosystem, even when they do not trust one another. As much as possible, we replace outdated recordings and inspections with IoT devices, effectively reducing human intervention in the system.

7.6.2 Blockchain for Smart Agriculture

Smart agriculture has made use of BC technology. The United Nations Food and Agriculture Organization (FAO) proposed that ICT e-agricultural infrastructure components converge BC technology requirements and ICT. They contend that threshold environmental and agricultural data integrity is safeguarded for those who engage in public data management when distributed ledger systems are used for record-keeping in immutable ICT e-agricultural systems. The study's writers (Patil et al. 2017) evaluated BC-based concepts linked with ICT-based technologies. Furthermore, they proposed an ICT e-agriculture framework with a BC architecture for local and regional use. The authors (Patil et al. 2017) presented a minimal BC-based design for secure and private smart greenhouse farms. Private immutable ledgers are advantageous for smart nodes in greenhouses that function as a BC managed centrally to enhance energy use. Additionally, they exhibited a security framework that combines IoT and BC technologies to offer a secure platform for smart agricultural techniques.

7.6.3 Intelligent Irrigation Driven by IoT

The authors (Cambra et al. 2017) offered a design for an intelligent IoT network communication system that might be employed as a cheap irrigation control device. The suggested irrigation tool uses real-time data, including field-measured parameters and variable rate irrigation. Stream threshold, pressure changes, wind velocity, benchmark vegetation (calculated utilizing aerial photographs), and irrigated agriculture activities are all regularly sampled. A clever cloud service processes data. For IoT smart irrigation systems, the authors (Kokkonis et al. 2017) suggested a new fuzzy mathematical approach. It lists every sensor, actuator, and microcontroller that might be used in water distribution systems. The irrigation system keeps an eye on the moisture content of the soil, humidity level, and temperature continuously. The field is covered with ground humidity sensors. The readings are sent to a microprocessor, which decides whether or when to open a servo valve based on a fuzzy computational method. The microcontroller sends all its data to a cloud database and then processes it for statistical analysis. The authors (Lin et al. 2018) also presented "Smart Irrigation Analysis," which is superior to conventional agricultural irrigation on the field in that it provides online analysis of irrigated agriculture on the farm to the end user. The intelligent irrigation program contains the following:

- An irrigation schedule that is automated and repeats.
- Data on the current moisture level
- Keeping track of and analyzing crop water use.

They use a microcontroller called the ESP8266 with a built-in Wi-Fi module. A hydration sensor is set up in the field to keep track of the amount of soil moisture. In order to make people's cultivating activities cheerful and more peaceful, the collected data are sent to the cloud. An irrigated agriculture graph report is created using cloud data after it has been evaluated to assist farmers in selecting the crop to plant in the future.

7.7 Application of BIoT in Healthcare

The IoT and BC are two examples of technologies that are already ubiquitous and moreover, applied in various fields, particularly e-healthcare. In healthcare, IoT devices can collect real-time patient sensory information, which can then be processed and evaluated. IoT data are collected, and centralized computing, processing, and storage are performed. Due to the potential for a single source of failure, privacy evasion, manipulation of data, interference, and suspicion, such centralized control can be risky. BC technology, which enables decentralized storage and processing of IoT data, has the potential to solve such pressing problems (Ray et al. 2021).

There are several business-related applications for BC technology. A fascinating application of BC technology is in the healthcare sector. By providing information on consumer expectations and preserving patient privacy, using the BC to pay fees with Bitcoin validates all stakeholders, including hospitals, healthcare providers, and health authorities. Data management, finance, cyber security, the IoT, food science, the healthcare industry, and brain research are just a few BC technology applications gaining increasing traction. There is much interest in applying BC technology to enable secure and private healthcare information management. BC is also changing conventional healthcare practices by facilitating secure, confidential information exchange and improved diagnosis and treatment. BC technology can develop customized, genuine, and safe healthcare in the future by combining all actual clinical data relevant to a patient's health and delivering it in a modern, secure healthcare environment (Siyal et al. 2019). The following are some ways that BC technology can change the healthcare industry

7.7.1 Interoperability

"The capacity of various software and information technology systems to interact, transfer information, and utilize the data that have been communicated is referred to as interoperability by the Health Information and Management Systems Society Gordon and Catalini (2018)." Interoperability in healthcare has several possible advantages. First, effective systems can increase operational effectiveness by cutting down on time devoted to administrative chores like manually inputting data obtained via faxes (Zhou et al. 2014). Additionally, interoperability can eliminate redundant clinical interventions such as imaging tests or lab requests, lowering the overall cost of the healthcare system, minimizing waste, and enhancing patient safety by lowering radiation and invasive procedure exposure (Walker et al. 2005; Stewart et al. 2010). The capacity to obtain pertinent, longitudinal clinical data more efficiently at the moment of care is another way interoperability may enhance clinical care.

7.7.2 Improved Analytics and Data Storage

Regarding visibility, audit, accountability, integrity, data authenticity, control systems, confidence, confidentiality, and safety, evolving healthcare information and management systems confront considerable hurdles. A natural disaster-related single point of breakdown is also more likely because most systems in use today for managing healthcare data are centralized. The way data are handled in the healthcare industry has the potential to change drastically. A new, innovative, and decentralized technology is BC. BC delivers an assurance of immutability and secure health data storage. BC can redefine and alter the medical industries by offering significant gains in efficiency, data protection, health professional staff management, and costs. Nevertheless, some technical issues need to be resolved when integrating healthcare systems with BC, including BC's lack of maturity, flexibility, interoperability, projects that stand alone, the difficulty of integrating them with current healthcare systems, their complexity, and a shortage of BC skill (Yaqoob et al. 2021).

7.7.3 Increased Security

The data-intensive healthcare sector generates, shares, stores, and retrieves a sizable amount of daily data. For instance, information is generated when a patient performs particular tests and must be given to the radiographer and, subsequently, a doctor. The session results are then recorded on file at the facility in case a doctor employed by another hospital in the network needs to see them later. Private and sensitive information can be found in healthcare data, which could be appealing to cybercriminals. As an illustration, cybercriminals who want to make money from the theft of these kinds of data may sell the information to an outside service, which might then analyze the data to find people who might not be eligible for insurance because of their medical records or genetic disease. Some organizations or industries would be interested in such data. Because of the interplay and intricacy between the components and systems, it is crucial to protect the security of the attacking ecosystem's principles related and elements and the medical ecosystem.

Furthermore, illegal access attempts from within the network or ecosystem must be prevented to protect the privacy and accuracy of medical data. Utilizing cryptographic primitives, such as those built on asymmetric key infrastructure and public clouds, to protect the secrecy and privacy of data. Data are encrypted, for instance, before being sent to the cloud. However, this restricts the data's searchability because healthcare practitioners first need to decrypt the data before looking for the decrypted data. As a result, data retrieval or diagnosis takes longer and costs more money (Esposito et al. 2018).

7.7.4 Immutability

BC technology offers an immutability feature that makes data and information unchangeable. The data are unchangeable when each patient has an unchangeable record, and accuracy is assured. Data analysis becomes simpler and almost error-proof for academics and medical professionals when information and data are more reliable. Additionally, immutability aids in record security and deters fraud and malfeasance.

7.7.5 Quicker Services

The healthcare sector benefits from Distributed Ledger Technology (DLT), which shortens patient and service provider wait times. The procedure between doctors, specialists, and health insurance providers is sped up when the information and data are accessible and exchanged on a distributed ledger, which speeds up the delivery of healthcare services to patients.

7.7.5.1 Transparency

Once all medical practitioners have equal access to up-to-date, unedited information, BC technology promotes transparency in the healthcare system. Health insurance, pharmaceuticals, healthcare research and innovation, private healthcare professionals, public healthcare systems, nursing homes, dentistry, and healthcare administration are just a few industries where BC applications could revolutionize and enhance the global healthcare delivery system.

7.8 Application of BIoT in Voting

Research on electronic voting systems has been ongoing for decades, intending to reduce election costs while maintaining election integrity by meeting compliance, security, and privacy standards. A novel election system that replaces the conventional pen-and-paper system may reduce fraud while improving voting process accountability and transparency (Hjálmarsson et al. 2018). By providing a fresh way around electronic voting methods' drawbacks and adoption difficulties, BC technology protects election integrity and security and paves the way for transparency. A BC-based electronic voting system employs smart contracts to safeguard voters' privacy while enabling secure and economic elections (Hjálmarsson et al. 2018).

IoT radically alters manufacturing and production in traditional organizations, and when integrated with BC, it can accomplish decentralized IoT. Decentralized IoT architecture can benefit from incorporating ledger self-tallying voting systems to address fairness concerns in self-tallying systems and ensure sturdy construction (Li et al. 2020). E-voting platforms could use BC technology, for example. Such a scheme would provide a decentralized architecture to administrate and support an open, equitable, and independently verifiable voting system. We can employ some effective electronic voting techniques that use the BC as an open voting system. The protocol must be created to adhere to crucial e-voting characteristics, provide some degree of decentralization, and allow voters to update their votes. Although the future of BC technology is very promising, its limitations may prevent it from realizing its full potential. Research on the fundamentals of BC technology needs to be intensified to enhance its capabilities and assistance for sophisticated apps that can use the BC ecosystem. Hardwick et al. (2018) shows that the privacy and safety concerns in e-voting systems create a big concern in which attackers can commit various frauds to rig the votes (Razzaq et al. 2017; Hjálmarsson et al. 2018; Iqbal et al. 2019). As a result, the main problem is distinguishing between ideal and malignant IoT devices to develop a genuine communication environment.

Furthermore, to prevent future changes to data gathered by smart devices, a BC is managed in which all genuine IoT device blocks are recorded (Chen et al. 2019: Lin et al. 2020). BC technology can address the growing demand for transparency, which is rising alarmingly. Furthermore, it can provide device security and transparency even if hackers hijack IoT items because any changes to the data can be immediately identified.

7.9 Application of BIoT in Supply Chain

It has been 13 years since the invention of Bitcoin and the BC concept. The initial intent was to create a peer-to-peer network-based resolution to the double-spending issue. BC has demonstrated its ability to bring a new level of trust to various services. Applications in government (land registries), electronics, and healthcare (patient records) are being investigated. The supply chain is one area in which BC technology is anticipated to be used. The report suggests using BC for such a supply chain by fusing it with distributed storage. BC is not a good option for data storage on a huge scale. It demands the storage of primary ledger data both on and off the BC.

Using accountability and incentive mechanisms to penalize and incentivize dishonest and trustworthy actors, respectively, is one way to improve data trust and reliability. These procedures depend on a trust management system, which we propose including in a supply chain management system. This type of system may also rely upon the data supplied by the IoT sensors (for example, temperature and location), which are increasingly being incorporated at various phases of the supply chain lifecycle (Haq and Esuka 2018; Malik et al. 2019).

IoT sensors should not, however, be relied upon blindly because they are vulnerable to errors or malicious attacks. Besides IoT sensors, other observations, like food authority review and approval and a seller's brand image in a food market, help build trust in the real world, and so on. Furthermore, conventional trust management techniques credit the image of the entities (Yang et al. 2017). It necessitates off-chain data preservations for intelligent contracts' verification and documentation and on-chain preservation of the fundamental ledger data. A practical approach is the Inter-Planetary File System (IPFS), a peer-to-peer decentralized file system that attempts to connect all computing systems with a single-file system. Large volumes of data can be addressed using the IPFS, and the unchangeable, permanent IPFS connections can be included in a BC transaction. Without placing the data onto the chain, this timestamp encrypts its content. Distributed storage and BC technology make the supply chain system suitable for the upcoming generation of industries. The BC-based system's features match those of Industry 4.0, and the model can help with these improvements (Kawaguchi 2019).

7.10 Summary

The popularity of cryptocurrencies and BC technology has grown over the last few years. Like how emails transmit information, they allow the transmission of value via the Internet. This means explicitly that peer-to-peer transactions can be conducted without financial intermediaries. Additionally, BC-based payment systems lack central organizations and are decentralized. Instead, they are governed by the group of people who utilize them. Decentralized payment systems can benefit people, businesses, and society because of their distinctive features, including reduced transaction costs and secure transactions. However, there are still some difficulties. Most importantly, decentralized payment systems lack firmly established institutions for people, and cryptocurrency exchange rates are highly volatile to trust.

This chapter's first section discussed BC technology and the IoT and the significance of BC technology applications. The second segment examined the history of BC technology applications, including what Bitcoin is, how it emerged, and how it works. This section also examined the numerous hazards and legal difficulties associated with Bitcoin. The third section highlights Hyperledger, Ethereum, and IOTA technology. The challenges and concerns associated with integrating IoT and BC are then discussed. The following section provided a more in-depth overview of many applications of this integration of BC technology with IoT (BIoT), like agriculture, healthcare, e-voting, and supply chain. The paper additionally looks at some challenges associated with implementing BC technology in the IoT market, such as scalability, privacy infringement, and security issues. It concludes by claiming that BC technology can transform the IoT business and open new avenues for innovation and growth.

We discovered that there are still areas of interest that need to be examined and explored in the future study. Establishing a standardized and interoperable network, privacy and security of IoT data, energy efficiency, regulatory compliance, and user adoption are among the future research areas for BC technology in IoT. To make BC-enabled IoT devices more energy-efficient, researchers must investigate novel encryption and authentication mechanisms, build new consensus algorithms, and investigate alternative energy sources. As BC technology becomes more widely used in the IoT industry, regulatory frameworks and compliance methods should be implemented to assure compliance with relevant legislation. Lastly, user attitudes and behavior toward BC technology must be examined to increase user adoption in a BC-enabled IoT ecosystem, and user-friendly interfaces and apps must be developed.

References

Alferidah, D.K. and Jhanjhi, N.Z. (2020). A review on security and privacy issues and challenges in internet of things. *International Journal of Computer Science and Network Security IJCSNS* 20: 263–286.

Alshaikhli, M., Tarek, E., Omar, E. et al. (2021). Evolution of internet of things from blockchain to IOTA: a survey. *IEEE Access* 10: 844–866.

Atlam, H.F., Alenezi, A., Alassafi, M.O., and Wills, G. (2018). Blockchain with internet of things: benefits, challenges, and future directions. *International Journal of Intelligent Systems and Applications* 10: 40–48.

Bahga, A. and Madisetti, V.K. (2016). Blockchain platform for industrial internet of things. *Journal of Software Engineering and Applications* 9: 533–546.

Barboutov, K., Furuskär, A., Inam, R., Lindberg, P. et al. (2017). *Ericsson Mobility Report*. Stockholm: Tech. Rep.

Biswas, K. and Muthukkumarasamy, V. (2016). Securing smart cities using blockchain technology, Sydney, NSW (12–14 December 2016), pp. 1392–1393. IEEE.

Boncea, R., Petre, I., Vevera, V. et al. (2019). Building trust among things in omniscient internet using blockchain technology. *Omanian Cyber Security Journal* 1: 17–24.

Buterin, V. (2014). A next-generation smart contract and decentralized application platform. *White Paper* 3: 2–1.

Cachin, C. (2016). Architecture of the hyperledger blockchain fabric. *Workshop on Distributed Cryptocurrencies and Consensus Ledgers* 310: 1–4.

Cambra, C., Sendra, S., Lloret, J., and Garcia, L. (2017). An IoT service-oriented system for agriculture monitoring, Paris (21–25 May 2017), pp. 1–6. IEEE.

Chen, J., Wu, J., Liang, H. et al. (2019). Collaborative trust blockchain based unbiased control transfer mechanism for industrial automation. *IEEE Transactions on Industry Applications* 56: 4478–4488.

Chilamkurti, N., Poongodi, T., and Balusamy, B. (2021). *Blockchain, Internet of Things, and Artificial Intelligence*. CRC Press.

Chod, J., Nikoloas, T., Gerry, T. et al. (2020). On the financing benefits of supply chain transparency and blockchain adoption. *Management Science* 66: 4378–4396.

Christidis, K. and Devetsikiotis, M. (2016). Blockchains and smart contracts for the internet of things. *IEEE Access* 4: 2292–2303.

Doguet, J.J. (2012). The nature of the form: legal ad regulatory issues surrounding the bitcoin digital currency system. *Louisiana Law Review* 73: 1119.

Doguet, J.J. (2013). The nature of the form: legal and regulatory issues surrounding the bitcoin digital currency system. *Louisiana Law Review* 73: 9.

Dorri, A., Kanhere, S.S., and Jurdak, R. (2016). Blockchain in internet of things: challenges and solutions. *arXiv preprint arXiv* 1608: 05187.

Esposito, C., De Santis, A., Tortora, G., Chang, H. et al. (2018). Blockchain: a panacea for healthcare cloud-based data security and privacy? *IEEE Cloud Computing* 5: 31–37.

Gordon, W.J. and Catalini, C. (2018). Blockchain technology for healthcare: facilitating the transition to patient-driven interoperability. *Computational and Structural Biotechnology Journal* 16: 224–230.

Grammatikis, P.I.R., Sarigiannidis, P.G., and Moscholios, I.D. (2019). Securing the internet of things: challenges, threats and solutions. *Internet of Things* 5: 41–70.

Gubbi, J., Buyya, R., Marusic, S., and Palaniswami, M. (2013). Internet of things (IoT): a vision, architectural elements, and future directions. *Future Generation Computer Systems* 29: 1645–1660.

Haq, I. and Esuka, O.M. (2018). Blockchain technology in pharmaceutical industry to prevent counterfeit drugs. *International Journal of Computer Applications* 180: 8–12.

Hardwick, F.S., Gioulis, A., Akram, R.N., and Markantonakis, K. (2018). E-voting with blockchain: an e-voting protocol with decentralisation and voter privacy, Halifax, NS (30 July 2018–3 August 2018), pp. 1561–1567. IEEE.

Hellani, H., Samhat, A.E., Chamoun, M. et al. (2018). On blockchain technology: overview of bitcoin and future insights, Beirut (14–16 November 2018), pp. 1–8. IEEE.

Hellani, H., Layth, S., Ben Hassine, M., Ellatif, A. et al. (2019). Tangle the blockchain: toward iota and blockchain integration for iot environment, Bhopal (10–12 December 2019), pp. 429–440.

Hjálmarsson, F.H., Hreiðarsson, G.K., Hamdaqa, M., and Hjálmtỳsson, G. (2018). Blockchain-based e-voting system, San Francisco, CA (2–7 July 2018), pp. 983–986. IEEE.

Hou, H. (2017). The application of blockchain technology in E-government in China, Vancouver, BC (31 July 2017–3 August 2017), pp. 1–4. IEEE.

Iqbal, R., Butt, T.A., Afzaal, M., and Salah, K. (2019). Trust management in social internet of vehicles: factors, challenges, blockchain, and fog solutions. *International Journal of Distributed Sensor Networks* 15: 1550147719825820.

Johnson, D., Menezes, A., and Vanstone, S. (2001). The elliptic curve digital signature algorithm (ECDSA). *International Journal of Information Security* 1: 36–63.

Kawaguchi, N. (2019). Application of blockchain to supply chain: flexible blockchain technology. *Procedia Computer Science* 164: 143–148.

Kokkonis, G., Kontogiannis, S., and Tomtsis, D. (2017). A smart IoT fuzzy irrigation system. *Power (mW)* 100: 25.

Kolias, C., Kambourakis, G., Stavrou, A., and Voas, J. (2017). DDoS in the IoT: mirai and other botnets. *Computer* 50: 80–84.

Konashevych, O. (2020). Constraints and benefits of the blockchain use for real estate and property rights. *Journal of Property, Planning and Environmental Law* 12: 109–127.

Krishnan, S., Emilia, V., Julie, E.G., Harold, R. et al. (2020). *Handbook of Research on Blockchain Technology*. Academic Press.

Li, X., Peng, J., Ting, C., Xiapu, L. et al. (2020). A survey on the security of blockchain systems. *Future Generation Computer Systems* 107: 841–853.

Li, Y., Susilo, W., Yang, G., Yu, Y. et al. (2020). A blockchain-based self-tallying voting protocol in decentralized IoT. *IEEE Transactions on Dependable and Secure Computing* 19: 119–130.

Lin, J., Shen, Z., Zhang, A., and Chai, Y. (2018). Blockchain and IoT based food traceability for smart agriculture, Singapore (28–31 July 2018), pp. 1–6.

Lin, X., Wu, J., Bashir, A.K., Li, J. et al. (2020). Blockchain-based incentive energy-knowledge trading in IoT: joint power transfer and AI design. *IEEE Internet of Things Journal* 9: 14685–14698.

Lu, Q. and Xu, X. (2017). Adaptable blockchain-based systems: a case study for product traceability. *IEEE Software* 34: 21–27.

Malik, S., Dedeoglu, V., Kanhere, S.S., and Jurdak, R. (2019). Trustchain: trust management in blockchain and iot supported supply chains, Atlanta, GA (14–17 July 2019), pp. 184–193. IEEE.

McCarthy, J., Otis, A., Sallie, E., Don, F. et al. (1800). *Situational Awareness*, 7B. *NIST Special Publication*.

Memon, M., Hussain, S.S., Bajwa, U.A., and Ikhlas, A. (2018). Blockchain beyond bitcoin: blockchain technology challenges and real-world applications, Southend (16–17 August 2018), pp. 29–34. IEEE.

Mistry, I., Tanwar, S., Tyagi, S., and Kumar, N. (2020). Blockchain for 5G-enabled IoT for industrial automation: a systematic review, solutions, and challenges. *Mechanical Systems and Signal Processing* 135: 106382.

Monrat, A.A., Schelén, O., and Andersson, K. (2019). A survey of blockchain from the perspectives of applications, challenges, and opportunities. *IEEE Access* 7: 117134–117151.

Panarello, A., Nachiket, T., Giovanni, M., Francesco, L. et al. (2018). Blockchain and iot integration: a systematic survey. *Sensors* 18: 2575.

Patil, A.S., Tama, B.A., Park, Y., and Rhee, K.-H. (2017). A framework for blockchain based secure smart green house farming. In: *Advances in Computer Science and Ubiquitous Computing*, 1162–1167. Springer.

Pustišek, M. and Kos, A. (2018). Approaches to front-end IoT application development for the ethereum blockchain. *Procedia Computer Science* 129: 410–419.

Rawat, A., Daza, V., and Signorini, M. (2022). Offline scaling of IoT devices in IOTA blockchain. *Sensors* 22: 1411.

Ray, P.P., Dash, D., Salah, K., and Kumar, N. (2021). Blockchain for IoT-based healthcare: background, consensus, platforms, and use cases. *IEEE Systems Journal* 15: 85–94.

Razzaq, M.A., Gill, S.H., Qureshi, M.A., and Ullah, S. (2017). Security issues in the internet of things (IoT): a comprehensive study. *International Journal of Advanced Computer Science and Applications* 8: https://doi.org/10.14569/IJACSA.2017.080650.

Reyna, A., Martin, Chen, J., Soler, E. et al. (2018). On blockchain and its integration with IoT. Challenges and opportunities. *Future Generation Computer Systems* 88: 173–190.

Samaniego, M. and Deters, R. (2016). Hosting virtual iot resources on edge-hosts with blockchain, Nadi (8–10 December 2016), pp. 116–119. IEEE.

Sarmah, S.S. (2018). Understanding blockchain technology. *Computer Science and Engineering* 8: 23–29.

Shammar, E.A. and Zahary, A.T. (2019). The internet of things (IoT): a survey of techniques, operating systems, and trends. *Library Hi Tech* 38: 5–66.

Sharma, G.D., Gupta, M., Mahendru, M., Bansal, S. et al. (2019). Emergence of bitcoin as an investment alternative: a systematic review and research agenda. *International Journal of Business & Information* 14: 3.

Shi, P., Wang, H., Yang, S., Chen, C. et al. (2021). Blockchain-based trusted data sharing among trusted stakeholders in IoT. *Software: Practice and Experience* 51: 2051–2064.

Siyal, A.A., Zahid, A., Zawish, M., Kainat, A. et al. (2019). Applications of blockchain technology in medicine and healthcare: challenges and future perspectives. *Cryptography* 3: 3.

Smith, B. and Christidis, K. (2016). IBM blockchain: an enterprise deployment of a distributed consensus-based transaction log, University of Alberta Edmonton, Canada (2–3 June 2016).

Sompolinsky, Y. and Zohar, A. (2015). Secure high-rate transaction processing in bitcoin, San Juan, Puerto Rico (26–30 January 2015), pp. 507–527. Springer Berlin Heidelberg.

Stewart, B.A., Fernandes, S., Rodriguez-Huertas, E., and Landzberg, M. (2010). A preliminary look at duplicate testing associated with lack of electronic health record interoperability for transferred patients. *Journal of the American Medical Informatics Association* 17: 341–344.

Tasatanattakool, P. and Techapanupreeda, C. (2018). Blockchain: challenges and applications, Chiang Mai (10–12 January 2018), pp. 473–475. IEEE.

Vivekanadam, B. (2020). Analysis of recent trend and applications in block chain technology. *Journal of ISMAC* 2: 200–206.

Vujičić, D., Jagodić, D., and Ranđić, S. (2018). Blockchain technology, bitcoin, and ethereum: a brief overview, East Sarajevo (21–23 March 2018), pp. 1–6. IEEE.

Walker, J., Eric, P., Douglas, J., Julia, A.-M. et al. (2005). The value of health care information exchange and interoperability: there is a business case to be made for spending money on a fully standardized nationwide system. *Health Affairs* 24: W5–W10.

Yan, Z., Zhang, P., and Vasilakos, A.V. (2014). A survey on trust management for Internet of Things. *Journal of Network and Computer Applications* 42: 120–134.

Yang, Z., Zheng, K., Yang, K., and Leung, V.C.M. (2017). A blockchain-based reputation system for data credibility assessment in vehicular networks, Montreal, QC (8–13 October 2017), pp. 1–5. IEEE.

Yaqoob, I., Salah, K., Jayaraman, R., and Al-Hammadi, Y. (2021). Blockchain for healthcare data management: opportunities, challenges, and future recommendations. *Neural Computing and Applications* 1–16. https://doi.org/10.1007/s00521-020-05519-w.

Zhang, Y. and Wen, J. (2015). An IoT electric business model based on the protocol of bitcoin, Paris (17–19 February 2015), pp. 184–191. IEEE.

Zheng, X., Zhu, Y., and Si, X. (2019). A survey on challenges and progresses in blockchain technologies: a performance and security perspective. *Applied Sciences* 9: 4731.

Zhidanov, K., Sergey, B., Alexandra, A., Mikhail, S. et al. (2019). Blockchain technology for smartphones and constrained IoT devices: a future perspective and implementation, Moscow (15–17 July 2019), pp. 20–27. IEEE.

Zhou, Y., Mandar, U., Sudeep, H. et al. (2014). The impact of interoperability of electronic health records on ambulatory physician practices: a discrete-event simulation study. *Journal of Innovation in Health Informatics* 21: 21–29.

8

Advanced Persistent Threat

Korean Cyber Security Knack Model Impost and Applicability

Indra Kumari[1,2] and Minho Lee[1,2]

[1] *Korea Institute of Science and Technology Information (KISTI), University of Science and Technology (UST), Daejeon, South Korea*
[2] *Department of Computer Engineering, Tongmyong University, Busan, South Korea*

8.1 Introduction

The Internet plays a very active role in every aspect of society in this age of technology. Without a doubt, it is evident that the Internet has a significant impact on both professional and personal experiences. Internet activity is irresponsible, just like the requirement for cover and evacuation. There is a clear distinction between intelligence that is kept private and data that may be gathered regarding data like phantasmagorias and scripts. The very same applies to affiliations that are either private or open. Also, levels and financial institutions now solely utilize the web for their daily operations through a declaration on online marketplaces. A much more dangerous aspect of the World Wide Web's widespread connectedness is its ability to prevent departure. Online security threats come from various sources, including cybercrime, spyware, hacking, and data theft. Over the ages, cybercriminals have developed cutting-edge, highly modern, and compelling ways to carry out these demands (Rao 2004). Advanced persistent threat, sometimes known as APT, is one of these powerful techniques used for infiltration. Advanced persistent threat attacks are generally referred to as APTs. APTs are online crimes consumed by both commercial and political sectors. For APTs to be effective, a decent concentration of secretive nature above a lengthy period is required. Even though the enemy's defenses have been breached and initial objectives have been exceeded, the assailant often aims for greater than rapid financial growth, and ill businesses are continued to be given up.

Machine Learning Applications: From Computer Vision to Robotics, First Edition.
Edited by Indranath Chatterjee and Sheetal Zalte.
© 2024 The Institute of Electrical and Electronics Engineers, Inc.
Published 2024 by John Wiley & Sons, Inc.

Computer technology and the quickly developing grid know-how are deceiving civilizations. These skills speed up computer hackers to the next level with harnessing skills. Cyberbattles are now more skilled, tenacious, and honed than ever before. An advanced threat attack is a novel type of cyberattack that has now been anticipated but also has gained popularity. Ingenious ideas, as well as excellent standards, must be kept in order to examine their complicated coordination during APT attacks. However, this might area might need to be clarified (Chaubey 2009). It is because APT attacks could well be thwarted by various links, such as interpersonal interconnections and computing networks, which function differently. One example could capture the complex interactions of APT in many patterns. Long-range or extensive APT attacks are another challenge. By simplifying conflicts as well as reducing exceptional complexity, renowned outstanding should be able to break such coincidences prototypically.

This article analyzes irregular humanitarian experiments conducted online while employing a basic harmonized rigorous strategy to manage real-world actions. The importance of deep ideological fortifications to governments and institutions cannot be overstated. People are prone to harmony despite the effectiveness of the increase in untrustworthy information, and the anthropomorphic component is the fragile relation (Rastogi and Regidi 2014).

8.2 Background Study

The phrase "advanced persistent threat" initially occurs inside the public sphere with a United States patent application. Increased complexity or expertise, quick teamwork, and progressively organized connections are characteristics of this strategy, which frequently overwhelms complicated security systems from within. Their strategy involves perseverance and secrecy, but its motive is gradually getting more income. This includes the potential government players for whom the actions support long-term manipulation and propaganda efforts, in addition to catastrophic results that support military intervention. These distinctive behaviors comprise the use of global agent systems, cyber malware and other sophisticated engineering tactics, protracted data analysis, and minimal attacks (Mouton et al. 2014).

The cybersecurity field did not just apply APT to government activities by the beginning of 2010. APT was a term used to describe lengthy cyber operations by well-funded opponents targeting to achieve specific goals employing advanced tactics, methods, and protocols (Mansfield 2014). Despite being commonly stated in this research, since APTs are carried out for influential organizations interested in the outcome, the consequences of such information are not thoroughly investigated. With varying levels of attention put into connecting such techniques in the framework within a specific operation or lifespan, most of the debate is focused on a strategic viewpoint on techniques.

The APT assault aims to remain undiscovered till the assault has been stopped, in addition to collecting information from a particular project. The well-funded hackers develop powerful equipment for all these purposes like novel varieties of ransomware seldom picked up by biometrics antivirus programs and detection and prevention software (Alavi et al. 2015). They collect all the data based on which they can rely on the company, including the strategies and tools it employs, the programs it serves, and the antivirus programs it uses. Additionally, they invest time in finding potential flaws within each program and developing spyware to take advantage of such flaws. Subsequently, in an effort to access an entire database, they disseminate these produced malware frequently through phishing and spear phishing attacks.

In order to fully understand APTs, it is worth reconsidering both fields of modern warfare and spatial awareness. First, APTs are deliberate, well-managed cyberattacks on a particular agency's digital products. Second, businesses must improve to make use of their contextual knowledge of APTs as well as the evolving security environment in order to battle APTs. Third, the art and science of using knowledge tactics ultimately determine the achievement of either side. A lengthy period of standard ways, research, creation, experimentation, application in practice, and experiences gained is presented in the military policy's accumulated, high-level results. APTs involve structured actions carried out with the assistance of goals that could extend far beyond the immediate effects of the strike on systems and data or the dependable harmful repercussions for people and organizations (Neupane et al. 2018).

Military science, which refers to the scientific knowledge of structured disputes, offers a valuable place for comprehending, specifying, and categorizing APTs. Such activities can potentially be critical strategies for revolution, allowing for the gradual acquisition of riches and power on behalf of many other individuals, groups, communities, and countries (Roohparvar 2019). APT actions are instances of planned warfare, wherein the purposeful use of assets is made by established groups to either assault or protect assets. Even though specific relevant laws must be completed in order for hostile activities to technically qualify as "acts of war," this type of war is often not acknowledged as being underneath such standards of worldwide law (Kethineni 2020).

The quick targets accomplished through every one of such activities contribute toward the accomplishment of longer-term organizational plans by APT controllers and any potential allies. For instance, while effectively attempting to steal the formula of an innovative product could perhaps reach the aim of the procedure designed to attack another corporation, effectively snatching a number of different types of trade secrets from a variety of different businesses in a similar industry could facilitate the accomplishment of a supporting energy's primary plan of monopolizing a particular economy (Dhengle and Nair 2022). The secret to stopping the APT is realizing that it is an organization. Activities involve mission endeavors that are constrained by the operator's specific number of tactical maneuvers for attaining one or more key goals. The operating vision establishes a connection between strategies and the accomplishment of operational targets that advance key goals. The defense may be capable of anticipating strategic actions, utilizing resources efficiently, or defeating the assault by attacking in the opposite direction as the assailant. Operations could be interfered with to prevent the enemy from reaching its goals, or they may even be altered (for example, by deliberate deception or deception) to thwart an enemy's larger organizational goals (Chetioui et al. 2022).

8.3 Literature Review

Nguyen (2016) postulated that cryptocurrencies like Bitcoin are shepherded on a viscount-to-patrician edifice network. Each earl has a comprehensive antiquity of all trades, thus showing the steadiness of each justification. This is rudimentary public-key cryptography and the edifice slab on which cryptocurrencies are based after being contracted, and the transaction is an announcement on the grid. When a viscount discerns a new transaction, it is patterned to ensure that the moniker is legal (Min 2019). If the corroboration is lawful, then the block is auxiliary to the chain; all other blocks further after it will "ratify" that transaction for cyber-bouts. Swan et al. (2017) recommended an intangible background to illuminate the collaboration between the Internet and cyber-bouts in emerging realms. The charter is established and constructed on the literature, circumstantial suggestion, and speculation. They have veteran extensive souk receiving and reckless improvement despite their contemporary notion. Many shrubbery doughs and aptitude bureaucrats have launched strategies to embrace cyber-allied money into their assortments and swap tactics. The speculative unrestricted has correspondingly disbursed sizable exertions in researching swap.

Dichotomized the adroitness and profitable swap panoramas in the advanced persistent threats cloister. Corroboration is a grave perception in cyber-bouts; only colliers can ratify contracts. Miners add chunks to the cyber-bouts; they salvage dealings in the preceding block, cartel it with the middle of the forgoing block to achieve its botch, and then accumulate the resulting hash into the current block (Pan et al. 2020).

Javed et al. (2022) Sappers in cyber-bouts receive transactions, smear them as genuine, and proclaim them transversely to the grid. After the driller ratifies the contract, each bulge must enhance its database. In layman tenures, it has become a chunk of the cyber-bouts, and miners undertake this work to achieve cyber threats. In contrast to cyber-security, they are allied to the procedure of tokens centered on disseminated ledger technology. Sternberg and Baruffaldi (2022) bequeathed an ephemeral appraisal on cyber-security associated to accord and equated diverse threats that are awaited. Cyber threats are the demonstrations used in these grids to propel assessment and emolument for these transactions. They can be alleged of as tackles on the health care and in some cases can also function as possessions or efficacies.

Gyimah et al. (2022) suggested that social engineering (SE) assaults can be dangerous. Their study creates a form of consciousness by classifying SE operations into two broad categories depending on their technique of incidence as well as the issues they cause. Additionally, they described the sources of information that may be used to start SE assaults and made several recommendations for mitigation variables that could be employed to limit the lower cyber likelihood of SE assaults on business targets. Everyone is susceptible to SE assaults; however, assailants concentrate on a specific person to gain entry to that person's organization. Once both technological and human aspects are considered, SE assaults can be carried out successfully. This study revealed that combating versus SE assaults should take into account both of the following: the formation of human consciousness via instruction and technological defense. (Chitrey et al. 2012) examined three attributes of assaults focused on social engineering: First, it makes it possible to portray social discipline assaults by creating a theoretical design. Second, it gives a way to gage how behavioral discipline operations affect individuals and institutions. Finally, it recommends using a protection technique to design a protection plan to thwart SE-based assaults. This requires a discussion of a study constraint. This study's researchers included students interested in data security, academics, and IT professionals in India. As a result, social networks are not seen from a non-IT standpoint.

Ivaturi et al. (2012) used the taxonomic method; their research attempted to clarify the various forms of SE assaults. They expect that by helping businesses fully appreciate known attacks, their taxonomy would help them develop robust and reliable defenses against the dangers they represent. There is little doubt that this categorization could be entirely thorough during this time; thus, they encourage additional change recommendations from the scholarly institution first. Interpersonal techniques are used to overcome technology defenses in a continuous interaction in which the aggressor would constantly strive to increase the stakes against the target if they are compelled to do so. As a result, they would continuously update their taxonomy that examines differences by recording them.

Social engineering attacks are becoming more and more dangerous. No security solution, no matter how sophisticated, can prevent social manipulation assaults. These assaults targeted not just the corporate company but also the broader population and clients of all providers. It is essential for both people and businesses to be conscious of various forms of social networking assaults and to follow or put into practice strategies for identification and abatement where applicable. Similarly, everyone must adopt certain data management practices to defend themselves from social engineering scams (Bhusal 2021). Federal agencies and network operators need to take a multimodal method toward prevention and damage mitigation. Tapscott and Tapscott (2022) bequeathed a fleeting appraisal of machine engineering spasm sorting. Any contract encompassing procurement, trade, and speculation encompasses an instinctive perfunctory. Healthcare is an ambitious framework and a solution that acts like a distributed network for the grid. The system crafts a revenue of transaction and facilitates the relocation of assessment and evidence. Healthcare is the gesture consumption in these networks to propel significance and remuneration for these transactions.

Aldawood and Skinner (2019) determined potential constraints created in implementing programs, regulations, or instruments. The authors assessed all multiple analyses for this realistic evaluation and examined the overall quality of techniques used, findings from the study, and conclusions. Learning, coaching, and raising awareness are used to prevent socially engineered assaults. Their evaluation additionally provides a general summary of how putting in place network security training and outreach initiatives might raise awareness in people. Finally, they aim to use this information to lessen and ultimately stop cyber defense social engineering assaults. Krombholz et al. (2015) discussed typical assault possibilities for contemporary cyber-attacks on skilled professionals in this research. Carrying its gadget regulations, dispersed cooperation, and contact through third-party networks all present new avenues of attack that may be used to carry out sophisticated social engineering assaults. They think a detailed overview of the security holes is required to create effective defenses to safeguard skilled professionals from malicious code assaults. Because of this, they developed a complete classification to categorize social engineering assaults based on the offensive route, its player, the various forms of social engineering, and certain assault circumstances.

Burda et al. (2021) made the case that various principles are urgently needed to defend against spear phishing assaults. More specifically, they suggested a novel plan of action using people security features like the foundation of automatic email methods to overcome the shortcomings of current defenses against this kind of threat. They are led chiefly by the belief that having a sizable percentage of people who have been resistant to hacking can benefit individuals who are not, thus increasing a company's core resistance. Initial findings indicate a future improvement in hacking information disclosure: measuring consumers' cognitive states. They illustrate the concept with a legitimate case and offer guidelines for making personal observations relevant. Weber et al. (2022) examine five instances involving cryptocurrency scams that significantly impact people. A comprehensive investigation of an incident is conducted utilizing an ethical platform for sustainable control assaults. Their study examines the psychological ploys or conformity techniques that the social engineers employed in various situations. By taking advantage of concepts like "Diversion," "Control," and "Dedication, Mutual support & Stability," the hackers were capable of accessing consumers' cryptocurrency-stored capital assets without damaging the cryptocurrency safeguards. The effectiveness of assailants is partly due to Bitcoin owners' ignorance of cybersecurity threats.

Postulated identifying security threats is one of the paramount solicitations in cyber-bouts for the credentials and verdict of problems and infirmities that are otherwise pondered stiff-to-perceive. This can embrace anything from the public system, stiff drumming to hook during the preliminary legs, to auxiliary maladies (Almeshekah and Spafford 2016). It is a crucial illustration of how assimilating discerning subtracting with security-centered soreness sequencing can comfort the fabrication of a debauched verdict. It opens the door for massive-to-ripen relaxing demeanors in capability. Ascertained discovery and manufacturing of security aspects is one of the decisive quantifiable solicitations of cyber threat learning defamations in a prompt-leg remedy detection manner. This also embraces expertise such as subsequent-cohort sequencing and meticulousness linctus Caglayan et al. (2012), which can provide coziness in verdict substitute trails for convalescence of multifactorial social maladies. Shortly, the cyber security learning enactments embroil hearsay erudition which can ascertain configurations in data without providing any extrapolations.

Ensley (2017) worked on numerous grounded cyber-bouts and securities blamable for the insurgency expertise plead computer apparition. This institute insurgency in the inward perceptiveness ingenuity is urbanized by Microsoft, in which drudgery on doppelgänger is indicative of contrivances for spitting image scrutiny. As APTs become complementary and convenient and proliferate in their illuminating dimensions, deducing the perception of superfluous data cradles from speckled remedial explanations becomes a chunk of this probing progression. Conceptualized modified linctus for bespoke enterprises can not only be auxiliary-operative by coupling explicit vigor with prophetic measures but is also apt for further exploration and enhanced security impost. Currently, empirical threats are inadequate to demonstrate from an explicit conventional mode of ascertaining or appraisal of the peril to the persistent fabricated on his reminiscent antiquity and manageable heritable evidence (Hutchins et al. 2011). However, APTs in linctus are abundant assembly gaits, and social sectors are at the vanguard of this effort by leveraging enduring social antiquity to engender numerous handling opportunities.

Maisey (2014) perceived the irrefutable probationary and considered that APTs have numerous impending solicitations in the pitch of the irrefutable probationary and probe, as anybody in the social diligence would articulate, and irrefutable provisional charge a lot of stretch and dough and can yield existences to widespread in sundry belongings. Smearing cyber-constructed prophetic pinpointing to ascertain impending quantifiable tentative contenders can help canvassers lure a puddle from an eclectic assortment of statistics. Cyber-bouts have also instituted convention in endorsing tangible-stint observing and data admittance of the probationary partakers, adjudging the preeminent illustration magnitude to be seasoned, and leveraging the clout of automated greatest to diminish data-based blunders. (Li et al. 2016) ascertained APT-grounded progressive amendment that paves that communicative amendment is an imperative chunk of preemptive linctus, and ever since the promulgation of cyber-bouts in social security with innumerable startups is garnering up in the pitches of assault deterrence and credentials, enduring controlling countenancing us to cognize our insentient comportment and sort indispensable fluctuations. The adverse effects of APT attacks were studied in their literature review, and a method was formulated to stop the attackers from getting access to the system.

APT attacks are a significant concern for organizations worldwide. APT attacks are sophisticated and targeted cyber-attacks that are designed to gain unauthorized access to a network or system and remain undetected for an extended period. APT attacks are typically carried out by well-funded and highly skilled groups, such as nation-state actors or criminal organizations.

According to a review of the literature on APT attacks, APT attacks are typically characterized by the following features:

1) **Targeted nature**: APT attacks are typically highly targeted and aimed at specific organizations or individuals. Attackers may gather information about the target organization and its employees through social engineering techniques, such as phishing or spear phishing, to gain access to sensitive data or gain employees' trust.
2) **Customized tools and payloads**: APT attacks often involve using custommade tools and payloads, such as malware or exploit code, tailored to the specific target. This helps increase the chances of success and makes it more difficult to detect the attack.
3) **Stealthy operation**: APT attacks are designed to remain undetected for an extended period. Attackers may use various techniques to avoid detection, such as hiding in legitimate traffic or using encrypted communications.

4) **Long-term objectives**: APT attacks are often carried out with long-term objectives, such as collecting sensitive data over time or gaining access to specific systems or networks.

Organizations can implement several measures to prevent and detect APT attacks, such as strong passwords and authentication protocols, regularly updating and patching systems, training employees to recognize and report suspicious activity, and implementing network and endpoint security solutions. It is also crucial for organizations to have a plan in place for responding to and recovering from a cyber incident. Overall, APT attacks pose a significant threat to organizations worldwide and require robust cybersecurity measures to detect and prevent them. It is essential for organizations to be aware of the characteristics and tactics of APT attacks and to implement effective countermeasures to protect against them.

8.4   Research Questions

1) What is the primary goal of APT?
2) What are the potential causes of said upsurge in APTs?

8.5   Research Objectives

The research's main principles are as follows:

1) To assess the effect of an APT in the Republic of Korea.
2) To investigate South Korea's knowledge and preventative actions regarding the APTt.

8.6   Research Hypothesis

H_0-1. There will be no substantial relationship between accessibility as well as the effect of APT in the Republic of Korea.
H_a-1. There is a considerable relationship between accessibility and lessening the effect of APT in the Republic of Korea.

8.7   Phases of APT Outbreak

APTs advance through five phases across duration to remain undetected. Here is a description of every stage of an active attack, spanning reconnaissance through information leakage (Figure 8.1).

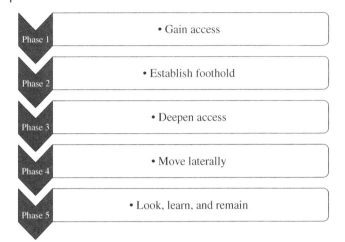

Figure 8.1 Phases of APT outbreak.

8.7.1 Gain Access

The attacker uses data from various sources throughout this early stage to understand the subject better. Due to their increased sophistication, cybercriminals can now target individual employees inside a business by collecting data through their web pages, profiles on social media, and other resources. Cybercriminals often enter a targeted system through a connection, an attached document, spam email, or a software weakness that injects a virus, much like a thief breaking open a doorway with a sledgehammer. Assailants may use Big Data processing and mining algorithms in addition to merely taking data from the web to evaluate the results and generate valuable insight continuously. Cyber attackers create an offensive approach or collect the required equipment depending on the information acquired. Hackers often arrange various instruments for various malicious pathways to get ahead so that they may change their strategies in case of problems. This phase is usually known as reconnaissance.

8.7.2 Establish Foothold

Hackers send a tailored virus to personnel or equipment that seems weak after gaining access to a victim's infrastructure. The APT is launched using a variety of malicious activities, such as phishing, social engineering, identity stealing, or push downloads. Hackers install a virus that enables the construction of a chain of vulnerabilities or gateways used to maneuver about in computers covertly. The virus frequently uses strategies involving program duplication to assist attackers and hide the traces. In the concentrated phishing technique known as spear phishing, chosen few receivers are the only ones who get the bogus emails. The idea of a watering hole assault is comparable to that of a predator lurking once at a waterhole in the wilderness because the attacker is aware that the prey must go to the drinking source. This phase is usually known as an incursion.

8.7.3 Deepen Access

Hackers act quietly and carefully throughout the investigation stage to prevent them from being discovered. Next, they develop an offensive approach, analyze a company's network-based protection measures, and set up many simultaneous assault routes, including one for web connection. Such simultaneous attacking paths provide a doorway into the perpetrator's IT system for potential data theft in the long term. The APT may increase access privileges throughout this stage to obtain control over information that is typically prohibited. They utilize methods like encryption and authentication to achieve admin privileges, and then they can take over more of the organization to increase accessibility. A backdoor trojan is often installed as a consequence of an effective infiltration. These cyber attackers then establish a connection with the victims' system. Data transmission is produced, and files are left behind on the victims' PCs, allowing defenses to identify an attack at an initial point.

8.7.4 Move Laterally

For more than a substantial amount of time, attackers allow entry to unsecured devices and gather information. The intruder has time to deploy spyware throughout this period since the target is unaware of an operation. To get critical corporate data, such as mail, papers, ideas, copyrighted material, or code editor, APT would inject maliciously. Once there, cybercriminals employ password-guessing strategies to achieve administrative privileges, which they could use to take over additional computer components and gain additional authority. Each phase often takes a long time since the assailants seek to collect as much data as possible across a considerable amount of time. APT attackers sometimes exploit genuine OS tools and features that IT managers frequently utilize, but they may additionally break or hijack passwords to get easy accessibility, making their actions invisible and maybe even undetectable.

8.7.5 Look, Learn, and Remain

During this point, an APT would pause unless there is an excellent chance to transfer recovered data back toward the scammer's control station for examination and possibly more deception and manipulation. Collected information may be encoded or delivered across hacked systems to conceal its destination and make it harder to recognize as hacked data. Within the infrastructure, attackers can fully comprehend how it functions and its points of vulnerability, making it possible to gather any data they choose. Attackers can continue this procedure forever or stop after achieving their objective. They generally leave a back entrance open to get into the network later. As APT attackers descend deeper, their activities are hard to track further into the system. Information leakage is a crucial stage for hackers because the main objective of an APT assault would be to collect sensitive information to obtain profitable opportunities.

8.8   Research Methodology

The research methodology for studying APT attacks will depend on the specific research question or objective being addressed. However, some common approaches that may be used include

1) **Data collection**: To study APT attacks, researchers may collect data from various sources, such as cyber incident reports, cybersecurity threat intelligence feeds, and open-source data. Researchers may also interview cybersecurity professionals or other experts to gather additional insights.
2) **Data analysis**: Once data have been collected, researchers will typically use various analytical techniques to identify patterns and trends in the data. This may involve using statistical analysis tools or conducting a qualitative analysis of text data.
3) **Case studies**: Researchers may also use case studies to examine specific APT attacks in detail. This may involve analyzing the tactics, techniques, and procedures used by the attackers and the responses and countermeasures implemented by the targeted organization.
4) **Simulation**: Researchers may also use simulation or modeling techniques to study APT attacks and the effectiveness of different countermeasures. This may involve building and testing various scenarios in a controlled environment to understand how different factors impact the likelihood and impact of an APT attack.
5) **Ethical considerations**: It is essential for researchers studying APT attacks to consider ethical issues, such as protecting the privacy of individuals and organizations and ensuring that research findings are not used for malicious purposes.

Overall, the research methodology for studying APT attacks will depend on the specific research question or objective being addressed, as well as the data and resources available. Researchers may use a combination of data collection, analysis, and simulation techniques to understand APT attacks and the effectiveness of different countermeasures.

For this paper, the campaign worker has presumptively used a theological method of investigation. The researcher has stopped using primary data based on a study of the research field and secondary sources like widely used literature on the research issue, such as books, artifacts from magazines, and correspondent discussions.

8.8.1 South Korea Cyber Security Initiatives and Applicability

The majority of South Korea's export markets, which are dominated by sectors focused on innovation, are machinery and equipment. The administration's top priority is the protection of advanced technologies due to the global financial significance of exporting and industry. The administration has focused its efforts on protecting the private, confidential data of companies that are deeply associated with exportation. Considering this official initiative, a large portion of academic organizations' safety work has focused on technology instruments to safeguard firms. Nearly three-quarters of research on corporate cybersecurity conducted between 2010 and 2016 has been reported to concentrate on technology security instruments. This characteristic also holds for studies of SMEs. Science research on SMEs' data protection centered on information leaks involving new technologies. The requirement for a coherent data protection policy could have been given more importance due to Korea's propensity to concentrate on technological improvement and the reality that various organizations are in charge of various parts of cyberspace. There has yet to be a lengthy strategy or elevated administrator capable of managing a successful national cyber program for years. However, based on the research and lessons learned from the previous assault, specific policy modifications were introduced following every operation to enhance incident management. South Korea has implemented several initiatives and policies to improve cybersecurity and protect against cyber threats. Some of these initiatives include

1) **Cyber security basic act**: This act, passed in 2015, establishes the framework for cybersecurity in South Korea and sets out the roles and responsibilities of government agencies and private sector organizations in protecting against cyber threats. It also outlines measures for responding to and recovering from cyber incidents.

2) **National cyber security center**: The National Cyber Security Center (NCSC) is a government agency coordinating cybersecurity efforts in South Korea. The NCSC works with other government agencies, academia, and the private sector to prevent, detect, and respond to cyber threats.

3) **Cybersecurity manpower development program**: This program, launched by the Ministry of Science and ICT, aims to enhance the skills and expertise of cybersecurity professionals in South Korea. It includes initiatives such as training programs, scholarships, and research grants.

4) **Cybersecurity information sharing platform**: The Cybersecurity Information Sharing Platform (CISP) is a system for sharing information about cyber threats and vulnerabilities among government agencies, critical infrastructure operators, and other stakeholders in South Korea. The CISP helps improve situational awareness and facilitate coordinated responses to cyber incidents.

These initiatives and policies apply to all organizations in South Korea, including government agencies, critical infrastructure operators, and private sector companies. They provide a framework for improving cybersecurity in the country and protecting against cyber threats.

The NIS encouraged the Basic Cyberwarfare Conduct, which also plays a significant part in securing national services, in 2016 once it was made aware of the issue. However, this was never passed because of disputes over the responsibilities and roles of governmental bodies and community mistrust of an intelligence organization. The administration did not create the initial Nationwide Security Plan before 2019. Eventually, the perpetrators of assaults received no reply. The purpose and history of assaults are assessed during coalition inquiries, and if North Korea is to blame, it is customary in Korea for that fact to be revealed to the public. In the eventuality of assaults from all other nations, the number of instances of open recognition will increase. In fact, in incidents unconnected to North Korea, the perpetrators have not yet been prosecuted or subject to independent penalties. For instance, South Korea has not made it known that perhaps the PyeongChang Olympics disaster was brought on by a Russian assault. Additionally, there have not been any instances of official exposure or punishments to assist in discouraging regular hacks emanating from China.

Through skillful deception, Korea intends to strengthen cyber retrenchment using fractious-edge cyberspace armament sequencing technology with global organizations, enabling obedient responses toward the threats posed by the world's virtual threats. The plan was developed using a local cybersecurity evaluation method that mimics Korea's unique and unique characteristics. Other nations may find it helpful to learn from South Korea's experiences when setting up procedures for preventing and responding to cyberattacks. Until recently, Korea followed similar sophisticated nations' cyberspace regulations, but it would be rapidly poised ever to become a recipient of numerous awards.

8.8.2 Korea's Cyber-Security Program Proposals

Korea espoused the solicitation of multi-proxy configurations to the delinquent of computerized indication throng and dispensation as well as the revealing of sneaky dogged menaces that cannot be effortlessly distinguished by decree-built structures that are as trails:

8.8.2.1 Modernized Multi-Negotiator Retreat Arrangement

The expansion of the rationalized multi-proxy retreat sorting targets to mend the tactic in which exposure is ended. The gigantic slant of checking all grid haulages and erecting a departing verdict genuine on an appraisal in contradiction of a moniker is inert and does not yield into justification alimental statistics about the mandate in which indication is institute or how it transmits to newfangled fragments of endorsement. The conventional architecture separates individual agent confirmation scrutiny and the comprehensive sorting moment to countenance for the inclusion of surplus contrivances such as proxy performance a vigorous classification for pivotal when the indication assortment juncture should be ruined, and the ordering accomplished.

8.8.2.2 Headway of the Realms Exemplary

The organization of purviews was engineering to drudgery with the discernment of disorders and belongings, staple to the maneuver of the dispersed multivalent retreat classification. The exemplary ascertains that expertise pledgee within indigenous grids transmit to apiece add in diverse approaches, and this is expenditure by the muggers to accomplish adjacent crusade around the web. The tagging increment hearings were conventional to sort expenditure of this conservational circumstance to arbiter the quality of the information found by the agents, where poised confirmation coined from around the grid in an erratic configuration, the probability of it actuality a contrived assenting upshot can be clinched, constructed on the classification of realms

8.8.2.3 Scrutiny of Over apt in Cyber Retreat

In retreat investigation, the concentrate of terminating proper usually verves unintentionally despite the milieu and figures customary being incredibly prone to its concurrent delinquents of creating erroneous examples that fail to acclimatize to new-fangled locales. Unlike other turfs of the amendment, muggers, predominantly progressive dogged menaces, have the goalmouth to endure veiled and betray those endeavoring to perceive them. Furthermore, muggers can unswervingly wobble the sorts' of connotation within the facts cliques used for drill and puzzling appliance erudition algorithms since the data pieces all instigate from moreover a manipulator or mugger. The appraisal illustrations that contrivance erudition algorithms often cripplingly partiality toward the precise documents conventional and the silhouette that the bout yields rather than the inclusive configuration of the bout.

8.8.2.4 Indiscriminate Inconsistency Revealing

The enrichment over the existing contrivance erudition tactics that frequently terminated stout to the mugger, the comprehensive glitch exposure algorithm was obtainable to perceive an eclectic assortment of spasms by cramming the adjacent crusade of assailants through the grid. The algorithm explicitly dodges consuming sorts of documents conventional and instead amends the recapitulations between assemblages of manipulators to treasure outliers for further consideration. The consequences illustrate that the algorithm can commence thriving in diverse milieus by puzzling it alongside diverse actualities conventional, comprehending an assortment of bouts. The detriment of the manufacturing algorithm is that it cannot ascertain the sort of bout but only the whereabouts of the bout for auxiliary ingravity exploration. The sweeping incongruity exposure algorithm and the scattered multi-proxy retreat classification drudgery are well-unruffled to distinguish the maiden smidgeons of a bout and then scrutinize its auxiliary. The indiscriminate glitch exposure algorithm discourses the feebleness intrinsic in the other classifications to afford a comprehensive elucidation to perceive and retort to cyber-bouts. The recompense of the indiscriminate incongruity exposure algorithm is its knack to-

1) Perceive an eclectic assortment of bouts, and
2) Maneuver on the grid echelon, while its feebleness is
3) The yield of the algorithm is a maneuver testimony that contrivances to be interim anomalously, not a comprehensive scrutiny of the occurrence.

The recompenses are as follows: -

1) Its knack to accomplish an in-profundity explorations into dealings,
2) Its propensity to pro-keenly wrinkle appropriate figures,
3) Its aptitude to exertion on-petition rather than as a visceral-potency exposure coordination, while its fragility is that the on-plea tactic entails a peripheral trace to unclutter the exploration. When used tranquil, the sorting discourses the delinquent of perceiving proceedings from a high layer.

8.9   A Deception Exemplary of Counter-Offensive

An austerely belligerent demeanor against APTs would necessitate potency prognosis yonder the garrisoned grid. Sweats would not only mark but also the machinists themselves, their arrangements, and all the ailments that empower their maneuver to embrace the knack and rheostat dimensions of any assistance individual transversely the virtual, perceptive, and somatic magnitudes of bustle subordinated with cipher feud. Legitimately discoursing such determinations is the purview of nation ceremonials that are authoritatively at confrontations are deliberately functioning within hoary stretches of transnational edict while steering astuteness undertakings, as thrilling as noncombatants and resident connotations are apprehension-offensive computer grids. There are endorsed techniques to flaccidly counteroffensive antagonists using falsehood (Figure 8.2). When machinists can be hoaxed into illuminating their maneuvers and their functioning goalmouths within a scrutinized milieu, it patronages the spread of pledge portions against them, thereby hypothetically terminating an existing maneuver while fabrication imminent maneuvers more challenging to mien.

Deception is accomplished accessible within the evidence milieu by the shielding adjacent. When these counterfeit rudiments are assimilated into an indulgent of the inclusive state on the aggressive adjacent, then delusion of the ailment befalls. This false impression twists the offensive horizontal's interpretation of connections between situational constituents, and it can consequence in fake extrapolation about how the circumstances will ripen in retort to assured intrusions. This indeterminate prognosis by the offensive adjacent in crack consequences in pronouncement to revenue a sequence of accomplishment which is instituted on the expectancy that does not copiously approve with the tangible locus. The belongings of the accomplishment can therefore diverge ominously from those awaited, and it is callous that the condition can then cultivate in unintentional tactics conceivably to the magnitude of "operation fiasco" for the confronting adjacent.

South Korea's technological growth has advanced quickly in the last two decades. State programs have succeeded in establishing nationwide infrastructures, enabling settings conducive to the World Wide Web. Because of this, the nation's

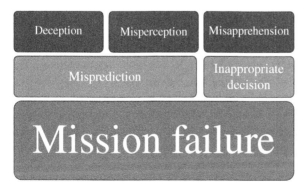

Figure 8.2   Stages of deception exemplary of counter offensive.

communications network is very good. Their broadband internet prevalence, or the number of subscribers for 100 residents, reached 41.2 in November 2017, placing it fifth internationally, and its mobile internet infiltration, at 111.7, placing them tenth worldwide. Almost all broadband internet users in South Korea possess links having rates faster over 10 Mb, allowing them to connect to the web quickly. This indicates that South Korea's average broadband rate is among the quickest in the world.

Cyberterrorism is viewed as the most significant security concern in South Korea. Cyber-terrorist strikes are an extremely efficient way for attackers to provoke and gain an advantage in future military operations. Defense specialists claim that hundreds of hackers are breaking into the internet defenses of potential prospective adversary countries daily. It is said to have targeted South Korea's public agencies, telecommunication hubs, and banking markets in six significant cyberterrorism attacks. These assaults caused South Korea to sustain significant harm. As a result, the government subsequently lost the people's trust through its capacity for controlling cyberspace. Due to attackers' online provocations, the leadership has been forced to concentrate its cyber defense strategy. This included ensuring security for critical national assets held by governmental organizations and significant enterprises instead of general cyber defense and ensuring the prevention of cyber warfare. Because of political factors surrounding the enemy, South Korea's security policy is now focused on combating cyberattacks instead of protecting single civilians and businesses. The State has recognized bank crimes as the only other danger to financial computer security, which is a serious societal issue.

The central governments and organizations in South Korea engaged in Cyber are classified into many categories based on the industry (army, commercial, and discreet). These departments and organizations, the Federal Spy Agency, the Intelligence Committee Federal Police Dept, the Ministry of Defense, and the Telecommunications Department of the Protection of Personal Information Commission Interiors, Education, Technology, and Strategic Development Ministries, are the organized departments working under state institutions. The Presidential Office's National Security Office serves as an operations center to address national cyberattacks, particularly contingency planning. Through directing the participating public and commercial domain organizations, the Cybersecurity Centre, part of the Intelligence Service, assists the Agency. The center does have the power to oversee a wide variety of operational activities, including legislative initiatives. According to governmental guidelines, all relevant authorities do their share to determine respective objectives. Therefore, the organization focuses on emergencies instead of prioritizing objectives or coordinating operations. There currently needs to be a regular procedure for evaluation inside the organization. It is claimed that no command employs a cyberspace security policy as a result.

Considering such coincidences toward its modern ideology style and supremacy, Korea noticeably lacks startled plot synopses in its cyberspace getaway doctrine. Unquestionably, reducing the use of Internet showcase the strengths and weaknesses of Korea's defensive cybersecurity abilities. The government needs to start by altering its behavior to deter and respond to impending cyber-battles. Even though Koreans have been the victim of numerous significant cyber-battles, a review of twenty centuries' worth of high-profile cyber-battles reveals little evidence of robust defenses on the part of the attackers. The Korean leadership is open about its propensity to enervate response to thefts that undermine local belief.

8.10   Conclusion

APT (Advanced Persistent Threat) attacks are sophisticated and targeted cyberattacks that are designed to gain unauthorized access to a network or system and remain undetected for an extended period. APTs have grown to be a serious threat to enterprises across many industries. There are no indications that this tendency will reverse. APTs are becoming an increasingly dangerous threat challenging to stop, identify, and fight against because of their stealth capabilities and reliance on data. Recently, there have been several instances when APTs have seriously harmed multiple companies, not just through security breaches but also severe violence. APTs use a sophisticated strategy to strike businesses; hackers are well-equipped and knowledgeable, and those who persist despite setbacks constantly hide their actual whereabouts and often focus on the precious commodities of the targets. Access control strategies are therefore required. Firewalls and virus protection are insufficient for corporations. Depending on these actions, they must use various sophisticated penetration detection systems and create new techniques for identifying abnormalities in a specific device and correlating them to identify security gaps. Although no system can be guaranteed to be 100% safe, companies may employ screening procedures that significantly lower the chance of an attempt. However, accomplishing that calls for expertise, materials, and commitment. APT attacks are typically carried out by well-funded and highly skilled groups, such as nation-state actors or criminal organizations.

To respond to APT attacks, organizations around the world have implemented several measures, including

1) Implementing strong passwords and authentication protocols helps prevent unauthorized access to networks and systems.
2) **Regularly updating and patching systems**: This helps address vulnerabilities that attackers could exploit.

3) **Training employees to recognize and report suspicious activity**: This helps detect and prevent APT attacks by increasing awareness of potential threats.

4) **Implementing network and endpoint security solutions**: This includes firewalls, intrusion detection, and prevention systems and antivirus software, which can help detect and prevent APT attacks.

5) **Establishing incident response plans**: This includes having a plan in place to respond to and recover from a cyber incident and regularly testing and reviewing the plan to ensure it is effective.

6) **Sharing information about threats and vulnerabilities**: This includes using cybersecurity information-sharing platforms and collaborating with other organizations and industry groups to share information about APT attacks and potential vulnerabilities.

Overall, organizations need to implement robust cybersecurity measures to detect and prevent APT attacks and to have a plan in place for responding to and recovering from a cyber incident if one does occur.

Acknowledgment

This research was supported by "Building a Data/AI-based problem solving system" of Korea Institute of Science and Technology Information (KISTI), and University of Science and Technology (UST), Daejeon, South Korea and my Professor, Min-ho Lee owe thanks for the support and excellent research facilities.

Conflict of Interest

The author declares that they have no known competing financial interests or personal relationships that could have appeared to influence the work reported in this book chapter.

References

Adu-Gyimah, S., Asante, G., and Boansi, O.K. (2022). Social engineering attacks: a clearer perspective. *International Journal of Computer Applications* 975: 8887.

Alavi, R., Islam, S., and Mouratidis, H. (2015). Human factors of social engineering attacks (SEAs) in hybrid cloud environment: threats and risks. *International Conference on Global Security, Safety, and Sustainability* (15–17 September, 2015), pp. 50–56. Cham: Springer.

Aldawood, H. and Skinner, G. (2019). A taxonomy for social engineering attacks via personal devices. *International Journal of Computer Applications* 176 (50): 19–26.

Almeshekah, M.H. and Spafford, E.H. (2016). Cyber security deception. In: Sushil Jajodia, V.S. Subrahmanian, Vipin Swarup, and Cliff Wang (Eds.), *Cyber Deception*, 23–50. Cham: Springer.

Bhusal, C.S. (2021). Systematic review on social engineering: hacking by manipulating humans. *Journal of Information Security* 12: 104–114.

Burda, P., Allodi, L., and Zannone, N. (2021). Dissecting social engineering attacks through the lenses of cognition. *2021 IEEE European Symposium on Security and Privacy Workshops (EuroS&PW)* (6–10 September, 2021), pp. 149–160. IEEE.

Caglayan, A., Toothaker, M., Drapeau, D. et al. (2012). Behavioral analysis of botnets for threat intelligence. *Information Systems and e-Business Management* 10 (4): 491–519.

Chaubey, R.K. (2009). *An Introduction to Cyber Crime and Cyber Law*. Kamal Law House.

Chetioui, K., Bah, B., Alami, A.O., and Bahnasse, A. (2022). Overview of social engineering attacks on social networks. *Procedia Computer Science* 198: 656–661.

Chitrey, A., Singh, D., and Singh, V. (2012). A comprehensive study of social engineering based attacks in India to develop a conceptual model. *International Journal of Information and Network Security* 1 (2): 45.

Dhengle, S.K. and Nair, N. (2022). Cyber crime and its laws in India as developing country. *International Journal of Law Management and Humanities* 5 (3): 1141.

Endsley, M.R. (2017). Toward a theory of situation awareness in dynamic systems. In: *Situational Awareness*, 9–42. Routledge.

Hutchins, E.M., Cloppert, M.J., and Amin, R.M. (2011). Intelligence-driven computer network defense informed by analysis of adversary campaigns and intrusion kill chains. *Leading Issues in Information Warfare & Security Research* 1 (1): 80.

Ivaturi, K. and Janczewski, L., 2012. A typology of social engineering attacks–an information science perspective. *AIS Electronics Library (AISeL)* 16 (4): 21–46.

Javed, S.H., Ahmad, M.B., Asif, M. et al. (2022). An intelligent system to detect advanced persistent threats in industrial internet of things (I-IoT). *Electronics* 11 (5): 742.

Kethineni, S. (2020). Cybercrime in India: laws, regulations, and enforcement mechanisms. In: Thomas J. Holt (Ed.), *The Palgrave Handbook of International Cybercrime and Cyberdeviance*, 305–326. Springer.

Krombholz, K., Hobel, H., Huber, M., and Weippl, E. (2015). Advanced social engineering attacks. *Journal of Information Security and Applications* 22: 113–122.

Li, M., Huang, W., Wang, Y. et al. (2016). The study of APT attack stage model. *2016 IEEE/ACIS 15th International Conference on Computer and Information Science (ICIS)* (26–29 June, 2016), pp. 1–5. IEEE.

Maisey, M. (2014). Moving to analysis-led cyber-security. *Network Security* 2014 (5): 5–12.

Mansfield-Devine, S. (2014). Hacking on an industrial scale. *Network Security* 2014 (9): 12–16.

Min, H. (2019). Blockchain technology for enhancing supply chain resilience. *Business Horizons* 62 (1): 35–45.

Mouton, F., Malan, M.M., Leenen, L., and Venter, H.S. (2014). Social engineering attack framework. *2014 Information Security for South Africa* (13–14 August, 2014), pp. 1–9. IEEE.

Neupane, A., Satvat, K., Saxena, N. et al. (2018). Do social disorders facilitate social engineering? A case study of autism and phishing attacks. *Proceedings of the 34th Annual Computer Security Applications Conference* (3–7 December, 2018), pp. 467–477. Association for Computing Machinery New York, NY, United States.

Nguyen, Q.K. (2016). Advanced persistent threat -a financial technology for future sustainable development. *2016 3rd International Conference on Green Technology and Sustainable Development (GTSD)* (24–25 November, 2016), pp. 51–54. IEEE: Kaohsiung, Taiwan.

Pan, X., Pan, X., Song, M. et al. (2020). Blockchain technology and enterprise operational capabilities: an empirical test. *International Journal of Information Management* 52: 101946.

Rao, J.S.V. (2004). *Law of Cyber Crimes and Information Technology Law*. New Delhi: Lexis-Nexis.

Rastogi, A. and Regidi, A. (2014). *Cyber Law: Law of Information Technology and Internet*. Lexis-Nexis.

Roohparvar, R. (2019). *Elements of Cyber Security*. InfoGuard Cyber Security.

Sternberg, H. and Baruffaldi, G. (2022). Advanced persistent threat in chains–logic and challenges of block chains in supply chains. *51st Hawaii International Conference on System Sciences* (4–7 January, 2022), pp. 3936–3943. Maui: Hawaii, USA.

Swan, M. (2017). Anticipating the economic benefits of advanced persistent threat. *Technology Innovation Management Review* 7 (10): 6–13.

Tapscott, A. and Tapscott, D. (2022). How advanced persistent threat is changing finance. *Harvard Business Review* 1 (9): 2–5.

Weber, K., Schütz, A.E., Fertig, T., and Müller, N.H. (2022). Exploiting the human factor: social engineering attacks on cryptocurrency users. *International Conference on Human-Computer Interaction* (26 June–1 July, 2022), pp. 650–668. Cham: Springer.

9

Integration of Blockchain Technology and Internet of Things: Challenges and Solutions

Aman Kumar Dhiman and Ajay Kumar

Department of Computer Science and Informatics, Central University of Himachal Pradesh, Dharamshala, Himachal Pradesh, India

9.1 Introduction

Blockchain is a distributed ledger technology that allows storing of data of digital transactions securely and openly. It uses cryptography to create a decentralized and tamper-proof system that does not require intermediaries, such as banks or other financial institutions. Each block in the blockchain contains a record of transactions that have been verified and added to the network by a consensus mechanism involving multiple participants. It creates a permanent and auditable record of all transactions on the network.

Internet of Things (IoT) refers to a network of actual physical objects, including cars, homes, and other items, that are connected to the Internet and can gather and share data. These gadgets have sensors, software, and other technologies built in that let them interact with each other and with other systems, such as cloud-based applications and databases. The information gathered by IoT devices may be utilized for many applications, including monitoring and controlling systems, improving efficiency, and providing new insights for businesses and consumers. However, the rapid growth of the IoT has also created significant challenges, especially in terms of security, privacy, and efficiency. Traditional centralized IoT systems are vulnerable to cyber-attack, data breaches, and single points of failure.

Machine Learning Applications: From Computer Vision to Robotics, First Edition.
Edited by Indranath Chatterjee and Sheetal Zalte.
© 2024 The Institute of Electrical and Electronics Engineers, Inc.
Published 2024 by John Wiley & Sons, Inc.

9.2 Overview of Blockchain–IoT Integration

Blockchain–IoT integration refers to merging blockchain technology and the IoT to create a secure, decentralized, and transparent network to exchange data and transactions between IoT devices. This integration can help solve several IoT challenges, such as data security, data privacy, and data sharing. The integration of blockchain and IoT enables the creation of a decentralized network of devices that can communicate and transact with each other without the need for intermediaries. Blockchain technology provides a secure and tamper-proof system for recording and verifying transactions and data exchanges. At the same time, IoT devices generate and transmit large amounts of data that can be used to trigger automated transactions and smart contracts. Some of the key benefits of blockchain–IoT integration include increased security and privacy of data, increased transparency, and accountability, and reduced costs and complexity of transactions. For example, in supply chain management, blockchain–IoT integration can help improve product traceability and authenticity, reduce the risk of fraud and counterfeiting, and enhance supply chain efficiency and transparency. Overall, blockchain–IoT integration is an emerging technology area with significant potential for innovation and impact on the industry and society. However, it also faces several challenges and limitations, such as scalability, interoperability, and standardization, which must be addressed to ensure its successful adoption and implementation.

9.3 How Blockchain–IoT Work Together

Blockchain and IoT are two distinct technologies that can work together to create new opportunities and applications. At a high level, blockchain provides a secure and decentralized platform for storing and sharing data, while IoT devices generate and transmit data to the network. Combining these technologies can create a new generation of secure and decentralized systems that can potentially transform many industries. Here are some ways in which blockchain and IoT work together:

1) **Immutable data storage and sharing**: One of the key benefits of blockchain is that it provides a secure and immutable platform for storing and sharing data. It is particularly useful in IoT, where devices and sensors generate large amounts of data. We can create a secure and transparent platform for storing and sharing IoT data with blockchain.

2) **Decentralized control and coordination**: Blockchain enables decentralized control and coordination of systems, which can be particularly useful in IoT.

By using blockchain to coordinate the actions of IoT devices, we can create a more efficient and resilient network. For example, in a smart grid system, blockchain could coordinate the actions of solar panels, batteries, and other devices to optimize energy production and consumption.

3) **Smart contracts and automation**: Blockchain networks can also support smart contracts, which have contract conditions between the buyer and seller encoded directly into lines of code and serve as self-executing contracts. Many procedures can be automated with smart contracts, such as payments, identity verification, and asset tracking. Smart contracts may be utilized in the IoT environment to automate device coordinating processes and control, creating more efficient and autonomous systems.

4) **Data privacy and security**: Blockchain can also provide enhanced privacy and security for IoT data. By using blockchain to store IoT data, we can ensure it is tamper-proof and secure from unauthorized access. Moreover, blockchain can establish secure and private data-sharing channels among devices, which is particularly important in healthcare and finance applications.

Overall, the combination of blockchain and IoT has the potential to create a new generation of secure and decentralized systems that can transform many industries. Using the strengths of both technologies, we can create more efficient, resilient, and autonomous systems that benefit the society.

9.3.1 Network in IoT Devices

A centralized network of IoT involves all devices being connected to a single, centralized server where all data collected by the devices are stored, managed, and accessed from a single location as shown in Figure 9.1. Connecting all IoT devices to this single server allows the server to monitor and manage data across all devices. When the server fails and attacks at one time, then all data are lost.

Figure 9.1 Centralized network.

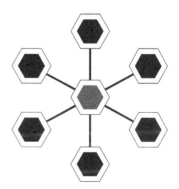

9.3.2 Network in IoT with Blockchain Technology

Blockchain technology works on decentralized (Figure 9.2) and distributed network (Figure 9.3). It has the potential to revolutionize the IoT industry. The blockchain allows IoT devices to securely share and store data, conduct transactions, and access services across different networks, regardless of the geographical location. As a result, IoT devices can collect and process data more efficiently, provide greater security and privacy for users, and reduce network transaction costs.

9.3.3 Data Flow in IoT Devices

IoT devices take in data through sensors and other inputs. A microcontroller processes these data and then sends them to a gateway access point. The data are then sent to a cloud platform where analytics and insights can occur as shown in Figure 9.4. Finally, the data are sent back to the IoT device in the form of control output to be used for various applications. When the centralized server fails and attacks at one time, then all data are lost.

Figure 9.2 Decentralized network.

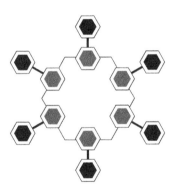

Figure 9.3 Distributed network.

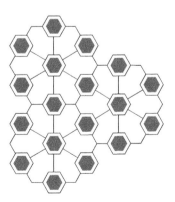

Figure 9.4 Data flow in IoT without blockchain.

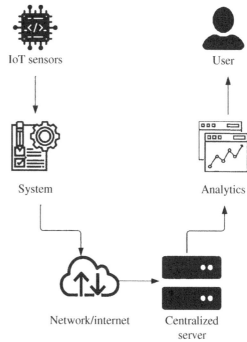

Figure 9.4 Data flow in IoT without blockchain.

9.3.4 Data Flow in IoT with Blockchain

Using blockchain technology in IoT devices enables secure, reliable, and efficient flow of data. It ensures that data are securely stored and only accessible to authorized users. Figure 9.5 shows data flow from IoT devices to the blockchain distributed ledger and are recorded with an immutable timestamp. As more users access the data, they are cryptographically secured via consensus protocols, making it almost impossible to change or lose data. Data can then be shared with other users with privacy and security at the top of their minds.

9.3.5 The Role of Blockchain in IoT

Blockchain can play a crucial role in integrating IoT by providing a secure and decentralized platform for storing and sharing data generated by IoT devices. Here are some specific ways in which blockchain can contribute to the IoT ecosystem:

1) **Data security**: Blockchain can provide a tamper-proof and immutable ledger for storing and sharing IoT data. The decentralized and distributed nature of the blockchain network makes it difficult for hackers to manipulate or corrupt data.
2) **Trust and transparency**: Blockchain can provide a transparent and trustworthy system for sharing data between IoT devices and users. Blockchain-based

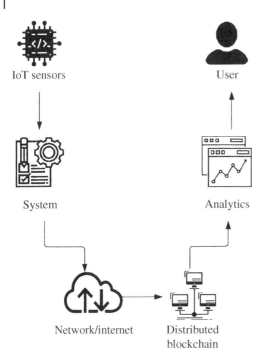

Figure 9.5 Data flow in IoT with blockchain.

solutions can ensure that the data are authentic, accurate, and have not been altered or tampered with.

3) **Smart contracts**: Blockchain can facilitate the execution of smart contracts, which are agreements that automatically execute, with the terms between the buyer and seller being directly included in the code. In the context of IoT, smart contracts can be used to automate and enforce the rules of data sharing, device control, and other processes.

4) **Data monetization**: Blockchain can enable the monetization of IoT data by providing a secure and transparent platform for selling and buying data. The decentralized nature of blockchain can eliminate intermediaries and ensure that data are sold at a fair price.

9.3.6 The Role of IoT in Blockchain

IoT devices can also play a crucial role in the blockchain ecosystem by generating data that can be used to validate transactions and provide more accurate information. Here are some specific ways in which IoT can contribute to the blockchain ecosystem:

1) **Data validation**: IoT devices can generate data that can be used to validate transactions on the blockchain network. For example, IoT devices can be used

to verify the authenticity and accuracy of supply chain data, such as tracking the movement of goods and verifying their origin.

2) **Decentralized data collection**: IoT devices can collect data in a decentralized and distributed manner, reducing reliance on centralized data centers. It can improve the speed and reliability of data collection, particularly in environments with limited Internet connectivity.

3) **Improved supply chain efficiency**: IoT devices can improve supply chain efficiency by tracking the movement of goods, monitoring inventory levels, and optimizing logistics. It can reduce costs and improve customer satisfaction.

9.4 Blockchain–IoT Applications

Blockchain technology has a wide range of applications and uses. When it comes to smart contracts, they are described as independent, decentralized bits of code that run when specific criteria are satisfied. Numerous real-world scenarios, such as foreign transfers, mortgages, or crowd fundraising, might benefit from smart contracts (permissioned distributed ledger).

Beyond cryptocurrencies and smart contracts, blockchain technologies can be used in a variety of fields (most of them are shown in Figure 9.6) where IoT

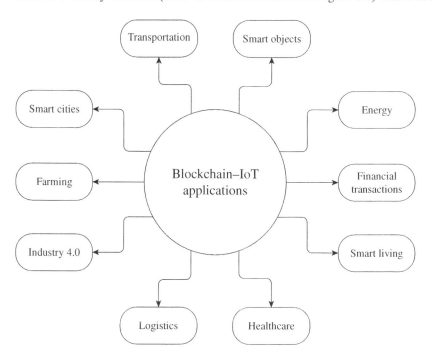

Figure 9.6 Blockchain–IoT applications.

applications are associated (Conoscenti et al. 2016), including sensing (Zhang and Wen 2015), data storage, identity management (Wilson and Ateniese 2015), time-stamping services, smart living applications, intelligent transportation systems, wearable sensors, supply chain management (Kshetri 2018), mobile crowdsensing, cyber law, and security in the IoT (Kshetri 2017).

Blockchain–IoT integration has several use cases in different industries. Here are some of the most prominent use cases:

1) **Supply chain management/logistics**: One of the most promising use cases of blockchain–IoT integration is in supply chain management. By integrating IoT sensors with blockchain, companies can track the movement of goods from the manufacturer to the end user in a tamper-proof and transparent manner. It can help reduce fraud, theft, and counterfeiting and improve the efficiency of the supply chain (Kim and Deka 2020).

2) **Smart homes and cities**: Another promising use case for blockchain–IoT integration is developing smart homes and cities. By integrating IoT devices with blockchain, cities can create a decentralized and autonomous system for managing resources like energy, water, and waste. It can reduce costs, improve efficiency, and increase the sustainability of cities (Kim and Deka 2020).

3) **Healthcare**: Blockchain–IoT integration has several potential use cases in healthcare. By integrating IoT devices with blockchain, healthcare providers can securely store and share patient data, track the movement of medical supplies and equipment, and improve the accuracy of diagnoses and treatment plans (Dwivedi et al. 2019).

4) **Energy management**: Blockchain–IoT integration can also improve energy management. By integrating IoT sensors with blockchain, energy providers can track energy consumption in real-time and adjust the supply accordingly. It can help reduce waste, improve efficiency, and lower consumer costs.

5) **Agriculture/farming**: Blockchain–IoT integration can also be used to improve agriculture by enabling the creation of a decentralized and transparent supply chain for agricultural products. By integrating IoT sensors with blockchain, farmers can track the growth and movement of crops from farm to table, reduce waste, and increase the quality of the produce (Kim and Deka 2020).

In summary, integrating blockchain and IoT can create new solutions and business models across different industries, leading to greater efficiency, security, and transparency.

9.5 Related Studies on Integration of IoT and Blockchain Applications

Table 9.1 Research papers on Blockchain–IoT integration.

Year	Article	Contributions
2023	Blockchain technology and internet of things: review, challenge and security concern.	This paper highlights that security is a major concern when it comes to IoT devices, which are vulnerable to hacking and other forms of cyberattacks. Blockchain can provide a layer of security by ensuring data integrity and preventing unauthorized access.
2021	Towards an interoperable blockchain based internet of things (Sajid et al. 2021)	In this paper, authors present a framework for the development of interoperable blockchain-based IoT systems, which involves the use of standard communication protocols, semantic modeling, and smart contracts.
2021	Towards blockchain-based secure, scalable, and resilient industrial internet of things(Cui et al. 2021)	The paper provides a valuable contribution to the field of blockchain-enabled IoT systems and highlights the potential of blockchain technology to address the challenges associated with IoT systems.
2021	A systematic review of blockchain-IoT integration research: a multiple case study approach (Wang et al. 2021)	The paper provides a detailed analysis of the multiple case studies, which demonstrate the potential of blockchain and IoT integration in various domains, such as healthcare, supply chain management, and energy management.
2020	Blockchain and internet of things integration: applications and challenge (Gao et al. 2020)	The authors provide a detailed overview of the various blockchain architectures and applications that can be used to integrate with IoT and highlight the need for continued research into the development of scalable, interoperable, and energy-efficient solutions.
2020	IoT and blockchain: a systematic review (Mohanty et al. 2020)	The paper provides a useful overview of the state of the art in IoT and blockchain research and highlights the potential of these technologies to transform various industries.
2019	A comprehensive survey on blockchain in IoT: applications, challenges, and opportunities (Li et al. 2019)	The paper highlights the challenges associated with integrating blockchain and IoT, including scalability, interoperability, and energy consumption. The authors provide insights into potential solutions for these challenges, such as off-chain solutions, sharding, and consensus mechanisms.

(Continued)

Table 9.1 (Continued)

Year	Article	Contributions
2019	Blockchain-based security for internet of things: a comprehensive survey (Suo et al. 2019)	The paper provides an overview of the various blockchain architectures that can be used in IoT systems, including public and private blockchain. The authors also describe the potential benefits of using blockchain in various IoT security applications, such as authentication, access control, and data integrity.
2019	A survey of blockchain in internet of things (Zhu et al. 2019)	The paper provides a comprehensive overview of the current state of blockchain applications in IoT systems, highlighting the potential of blockchain technology to address the challenges associated with IoT security and trust.
2019	A review of blockchain integration with IoT: challenges and solutions" (Nguyen et al. 2019)	In this paper, the authors discuss the challenges of security and privacy in the context of blockchain and IoT integration. They explore various approaches to addressing these challenges, including the use of encryption, access control, and smart contracts.
2019	A Review of Blockchain Integration with IoT: Challenges and Solutions (Nguyen et al. 2019)	The paper provides a valuable resource for anyone interested in understanding the current state and future potential of blockchain-based IoT applications and the challenges and solutions that must be addressed to fully realize the benefits of this technology. The authors also describe the potential benefits of using blockchain in various IoT applications, such as supply chain management, smart homes, and healthcare.
2019	Blockchain for Secure and Efficient Data Sharing in IoT: A Comprehensive Survey (Khan et al. 2019)	The paper also discusses different types of blockchain networks and consensus algorithms, as well as the scalability and interoperability issues that need to be addressed for blockchain to be more widely adopted in the IoT industry
2019	Blockchain and IoT: a systematic review (Mohanty et al. 2019)	The paper presents a detailed taxonomy of the different approaches to integrating blockchain and IoT and provides a comparative analysis of the various blockchain-based solutions proposed for IoT.
2018	Blockchain-enabled secure and efficient data sharing for industrial internet of things (Gao et al. 2018)	This paper provides a valuable contribution to the field of blockchain-enabled data sharing in the IoT and highlights the potential of blockchain to address the challenges associated with data sharing.

Table 9.1 (Continued)

Year	Article	Contributions
2018	Blockchain for IoT security and privacy: the case study of a smart home (Sicari et al. 2018)	The paper provides a useful example of how blockchain can be used to enhance security and privacy of IoT systems and highlights the potential of this technology to address the growing security and privacy concerns associated with the increasing use of IoT devices in our daily lives.
2018	A review of the use of blockchain in the internet of things" Lane et al. (2018)	The authors begin by describing the basic concepts of blockchain and how it can be applied to IoT systems. They discuss the advantages and challenges of using blockchain in IoT, including issues related to security, scalability, and interoperability.

9.6   Challenges of Blockchain–IoT Integration

1) **Scalability**: IoT devices generate vast amounts of data that must be stored and processed. Blockchain technology, on the other hand, has limitations in terms of scalability. The current blockchain infrastructure cannot handle the volume of data generated by IoT devices, resulting in slow transaction processing times (Reyna et al. 2018).
2) **Interoperability**: IoT devices operate on different platforms and protocols, making it challenging to establish a standard interface for communication. Blockchain technology also operates on different platforms and protocols, making it difficult to achieve interoperability between the two technologies (Farahani et al. 2021).
3) **Security**: IoT devices are vulnerable to security breaches due to their interconnected nature. Blockchain technology provides a secure and immutable ledger but is not immune to attacks. The integration of blockchain and IoT must ensure that security vulnerabilities are addressed (Hassan et al. 2019).
4) **Privacy**: IoT devices collect personal data, and the integration of blockchain technology could potentially expose these data to the public. Ensuring privacy and data protection is critical for successful integration (Hassan et al. 2019).

9.7   Solutions of Blockchain-IoT Integration

1) **Scalability solutions**: The use of sharding and sidechains can help increase the scalability of blockchain technology. Sharding involves dividing the blockchain into smaller parts, allowing faster transaction processing times.

Sidechains allow for the creation of parallel blockchain networks that can process transactions independently of the main network (Reyna et al. 2018).

2) **Interoperability solutions**: The use of standard protocols that can help achieve interoperability between IoT devices. Adopting common blockchain standards such as ERC-20 and Hyperledger Fabric can also help achieve interoperability between blockchain networks. (Farahani et al. 2021).

3) **Security solutions**: Using multi-factor authentication and encryption can help ensure the security of IoT devices. Integrating blockchain technology can also provide a more secure and decentralized infrastructure for IoT devices. (Hassan et al. 2019).

4) **Privacy solutions**: Using privacy-preserving techniques such as zero-knowledge proofs can help ensure the privacy of IoT data. Integrating blockchain technology can also provide a secure and decentralized infrastructure for managing personal data. (Hassan et al. 2019).

9.8 Future Directions for Blockchain–IoT Integration

It is an area of technology that is rapidly evolving and holds great potential for future development. Here are some of the key future directions for this technology:

1) **Emerging technologies and trends**: As blockchain and IoT mature, they will likely be integrated with other emerging technologies and trends, such as artificial intelligence, machine learning, and edge computing. It will enable new use cases and applications that are currently not possible.

2) **Opportunities for innovation**: Integrating blockchain and IoT creates a fertile ground for innovation in many areas, such as cybersecurity, data privacy, and decentralized systems. Companies and organizations investing in blockchain–IoT integration will likely be at the forefront of innovation in their industries.

3) **Potential impact on society**: Integrating blockchain and IoT can significantly impact society by enabling new business models, improving supply chain efficiency, enhancing cybersecurity, and increasing transparency and accountability. However, ensuring that the technology is used responsibly and ethically is important.

Blockchain–IoT integration is a rapidly evolving technology with great potential for future development. As emerging technologies and trends continue to emerge, there will be new opportunities for innovation in this field. The impact of blockchain–IoT integration on society will be significant, and it is important to ensure that the technology is used responsibly and ethically.

9.9 Conclusion

In conclusion, integrating blockchain and IoT represents a promising technology area with significant potential for innovation and impact on industry and society. By combining the security, transparency, and decentralized nature of blockchain with IoT's connectivity and data generation capabilities, companies and organizations can create new solutions and business models that were previously not possible. In this chapter, we have discussed the role of blockchain in IoT and the role of IoT in blockchain. We have also highlighted the advantages and challenges of blockchain–IoT integration and several use cases for this technology in different industries. Integrating blockchain and IoT can significantly impact society by enabling new business models, improving supply chain efficiency, enhancing cybersecurity, and increasing transparency and accountability. However, ensuring that the technology is used responsibly and ethically is also important. We believe there is a need for greater investment in research and development in this area and greater collaboration between industry, academia, and government. It will make it easier to guarantee that the technology is created morally and responsibly and that everyone may profit from it. In summary, the integration of blockchain and IoT represents a new era of decentralized and autonomous systems that have the potential to significantly impact industry and society. Our collective responsibility is to ensure that this technology is developed responsibly and ethically and that everyone benefits from it.

References

Conoscenti, M., Vetro, A., and De Martin, J.C. (2016). Blockchain for the internet of things: a systematic literature review. *2016 IEEE/ACS 13th International Conference of Computer Systems and Applications (AICCSA)*, (29 November–2 December, 2016), pp. 1–6. IEEE.

Cui, P., Wu, H., Wang, X., and He, Q. (2021). Towards blockchain-based secure, scalable, and resilient industrial internet of things. *IEEE Internet of Things Journal* 8 (9): 7469–7482.

Dwivedi, A.D., Srivastava, G., Dhar, S., and Singh, R. (2019). A decentralized privacy-preserving healthcare blockchain for IoT. *Sensors* 19 (2): 326.

Farahani, B., Firouzi, F., and Luecking, M. (2021). The convergence of IoT and distributed ledger technologies (dlt): opportunities, challenges, and solutions. *Journal of Network and Computer Applications* 177: 102936.

Gao, P., Zhang, X., Yang, L.T., and Zhou, Y. (2018). Blockchain-enabled secure and efficient data sharing for industrial internet of things. *IEEE Transactions on Industrial Informatics* 14 (11): 4987–4998.

Gao, H., Dai, X., Li, H., and Wen, Q. (2020). Blockchain and internet of things integration: applications and challenges. *IEEE Access* 8: 46211–46228.

Hassan, M.U., Rehmani, M.H., and Chen, J. (2019). Privacy preservation in blockchain based IoT systems: integration issues, prospects, challenges, and future research directions. *Future Generation Computer Systems* 97: 512–529.

Khan, M., Salah, A., Alghazzawi, M. et al. (2019). Blockchain for secure and efficient data sharing in IoT: a comprehensive survey. *IEEE Access* 7: 45101–45123.

Kim, S. and Deka, G.C. (2020). *Advanced Applications of Blockchain Technology*. Springer.

Kshetri, N. (2017). Blockchain's roles in strengthening cybersecurity and protecting privacy. *Telecommunications Policy* 41 (10): 1027–1038.

Kshetri, N. (2018). 1 blockchain's roles in meeting key supply chain management objectives. *International Journal of Information Management* 39: 80–89.

Lane, N.D., Dhillon, D., and Alani, A. (2018). A review of the use of blockchain in the internet of things. *IEEE Internet of Things Journal* 6 (2): 2441–2454.

Li, X., Jiang, P., Chen, T. et al. (2019). A comprehensive survey on blockchain in internet of things: applications, challenges, and opportunities. *IEEE Internet of Things Journal* 6 (5): 4177–4196.

Mohanty, S.P., Lenka, S.K., and Panda, B. (2019). Blockchain and IoT: a systematic review. *IEEE Transactions on Engineering Management* 66 (4): 934–946.

Mohanty, S.P., Lenka, S.K., and Panda, B. (2020). Iot and blockchain: a systematic review. *IEEE Internet of Things Journal* 7 (12): 12291–12313.

Nguyen, T.H., Nguyen, N.T., and Nguyen, T.T. (2019). A review of blockchain integration with IoT: challenges and solutions. *Journal of Computer Science and Cybernetics* 35 (4): 343–359.

Reyna, A., Martín, C., Chen, J. et al. (2018). On blockchain and its integration with IoT. Challenges and opportunities. *Future Generation Computer Systems* 88: 173–190.

Sajid, M.R., Abbas, H., Zhang, Y., and Madani, S.A. (2021). Towards an interoperable blockchain-based internet of things. *IEEE Internet of Things Journal* 8 (6): 4447–4463.

Sicari, S., Rizzardi, A., and Grieco, L.A. (2018). Blockchain for IoT security and privacy: the case study of a smart home. *IEEE Internet of Things Journal* 5 (6): 4305–4315.

Suo, H., Zhu, Q., Wu, Y., and Chen, Y. (2019). Blockchain-based security for internet of things: a comprehensive survey. *IEEE Access* 7: 10127–10141.

Wang, Q., Liu, X., Chen, Z. et al. (2021). A systematic review of blockchain-IoT integration research: a multiple case study approach. *IEEE Internet of Things Journal* 8 (7): 5342–5358.

Wilson, D. and Ateniese, G. (2015). From pretty good to great: enhancing pgp using bitcoin and the blockchain. *Network and System Security: 9th International*

Conference, NSS 2015, New York, USA (29 November–2 December, 2016), pp. 368–375. Springer.

Zhang, Y. and Wen, J. (2015). An IoT electric business model based on the protocol of bitcoin. *2015 18th International Conference on Intelligence in Next Generation Networks* (29 November–2 December, 2016), pp. 184–191. IEEE.

Zhang, J., Xu, Z., Chen, X., and Wang, C. (2019). A survey on blockchain-based IoT applications: challenges, solutions, and future directions. *IEEE Access* 7: 159436–159457.

Zhu, X., Wang, X., Chen, T., and Chen, X. (2019). A survey of blockchain in internet of things. *IEEE Access* 7: 82065–82081.

10

Machine Learning Techniques for SWOT Analysis of Online Education System

Priyanka P. Shinde[1], Varsha P. Desai[2], T. Ganesh Kumar[3], Kavita S. Oza[4], and Sheetal Zalte-Gaikwad[5]

[1] Government College of Engineering, Karad, affiliated to Shivaji University, Kolhapur, Maharashtra, India
[2] School of Computing Science and Engineering, D.Y.Patil Agricultural & Technical University, Kolhapur, Maharashtra, India
[3] School of Computing Science and Engineering, Galgotias University, Delhi, New Delhi, India
[4,5] Computer Science Department, Shivaji University, Kolhapur, Maharashtra, India

10.1   Introduction

The Indian government took many measures to stop spreading of COVID-19. The different measures like lockdown, curfews, self-quarantines, bans on public gathering, social distancing, and travel restriction were implemented. The COVID-19 epidemic affected daily life around the globe in 2019, having an effect on a variety of societal sectors including the private and public health among many other industries and the teaching profession. As a result, the educational system was suddenly terminated, and e-learning had to take its place as the mode of delivery. However, both students and faculty were unprepared for this abrupt transformation and were forced to quickly adjust and find ways to continue teaching in spite of a number of difficulties. Additionally, the epidemic had an impact on people's everyday lives because of loneliness, which led to anxiety and depression in many. Although the disruptions to the education sector created certain challenges, (Hegazi et al. 2022) it also increased awareness of the use of online technology in regular classes.

Both colleges and students can benefit from online learning. The students prefer flexible time for studying in developed nations like Australia and Korea because it made it easier for them to access the teaching materials. The adoption of this teaching strategy, which has a wider market and is more cost-effective due to its lack of facilities, also has benefits for universities. However, there are certain challenges (Chaturvedi et al. 2021) with online instruction because it takes substantial time for instructors to prepare the materials into the online version.

COVID-19 has become a vast topic of discussion. The entire world has faced this pandemic and still some of symptoms are active. Due to this pandemic, various fields got impacted, even we can say each and everything got impacted. The education system got affected as all schools were closed. There was lockdown in India, which caused schools and colleges to shut-down. After sometime, the online education system came into existence. Educational institutions were shut-down. All classrooms were closed and replaced by online classes. However, many students faced many difficulties to access online classes due to lack of Internet access, laptops/PC/mobiles, etc. In this pandemic, the development of online courses was on a large scale. Due to the lockdown, all colleges and schools were closed. The requirement increases, but with lack of awareness about using this technology for their daily education. Mostly, it was one of the major problems observed in the rural area. The study has been carried out to analyse educational needs before and after pandemic. But as online education started, we could not implement technology practically. There is a big lack of technology knowledge. The main disadvantage is the communication gap between students and teachers. This system also created a lack of good communication skills (Schnell and Krampe 2020) and due to which there is also lack of confidence in students to face offline classes.

Although online education has several drawbacks and shortcomings, it was a blessing in this epidemic. E-learning is expensive and time-consuming. It is not as simple as it seems. Significant expenditure is needed for the acquisition and maintenance of equipment, the hiring of skilled labor, and the creation of online contents. To implement education through an online manner, a powerful educational system must be created. Making sure of digital equity is essential in these tough times. Not every user has access to every digital resource, including WiFi, (Gournellis and Efstathiou 2021) the Internet, and digital tools. The goal of institutions should be to guarantee that all faculty members and students get access to the tools they require.

10.2 Motivation

COVID-19 has become a vast topic of discussion. The entire world has faced this pandemic, and still some of the symptoms are active. Due to this pandemic, various fields got impacted. Even we can say that everything got impacted. The major issue was in the educational field as well. The aim of the research paper is to propose a continuously tracking system on the impact caused on students during the online and offline education system. The system provides a detection where health issues and many other causative reasons are observed. The difficulty to attend the online lectures and unavailable mode of source also made students to face challenges. A sustained e-learning program is thus required (Ayuso-Mateos et al. 2021) since mentoring programs never succeed in life without being led by a broader conception.

The study aims to predict the impact caused on students and which can be cured by carrying out analyses for better future of students. For this research, we have studied different research papers and the dataset. In this, the different points of view of students have been considered and which impact caused is detected by using various algorithms. The study further states whether which online education system will you prefer for the further future scope of students. After obtaining the accuracy, (Papadopoulou et al. 2021) we can state whether a student has witnessed the impact of COVID-19 or on its education system. This will give a great and broad analysis of a student's brief education which he/she faced online as well as offline.

10.3 Objectives

1) To study the impact of COVID-19 on students' attentiveness in the online education system.
2) Implement machine learning algorithms to predict the impact of COVID-19 on online education and determine the best fitted model.

10.4 Methodology

The goal of this paper is to detect the post COVID-19 impact on school and college students' education system. School and college students faced different types of difficulties for a variety of reasons, including mental health issues, physical issues, and difficulty in understanding online education. The whole analysis is performed in the python environment (Jupyter Notebook). The following figure shows the work done from starting for calculating the accuracy of different algorithms.

As a result of our analysis, survey information was gathered by developing a questionnaire with the use of a Google form that is focused on aspects in order to ascertain the amount of impact on the education system of students. Students from various schools and institutions were asked to complete a form with personal data, which is accurate. The dataset consists of structured data in CSV

Figure 10.1 Diagram of the process of methodology.

format. The precise time the student entered his or her information is recorded in the dataset's timestamp. The personal information includes Full Name, Age, Gender, etc. Different attribute questions like facing challenges during online classes, having health issues post offline classes, knowing the platform used for entertainment purposes, different sites used for self-learning, and communication gap between students and teachers during online education were analyzed, and some were ranked accordingly on a scale from 0 (not very well) to 5 (very well), which defined the point of view experienced by students. (Hsu and Hsu 2022) However, some people were so self-conscious that they left some fields blank. The given framework supports to achieve the above set of the objectives-

1) First we make the best questionnaires using the different attributes with the help of a Google form which is helpful to detect the impact level on students.
2) After collecting the data, we can pre-process the given data.
3) When we complete data pre-processing, we can select the data which are helpful to apply to different algorithms.
4) There are six machine learning algorithms that were applied on the given data: K-Nearest Neighbor (KNN), Naïve Bayes (NB), Decision Tree (DT), Random Forest (RF), Linear Regression (LR), and Support Vector Machine (SVM).
5) After applying different algorithms (Bhatt et al. 2022), we estimated the accuracy level of each algorithm.

Figure 10.1 depicts the steps performed for analysis and prediction of the impact of COVID-19 on the education system.

10.5 Dataset Preparation

The Google form surveys students of different colleges to collect the data. This Google form consists of different attributes/questions to provide information related to their education system. The dataset consists of 650 rows and 55 columns.

Table 10.1. Consists of all the attributes of the dataset with its data type. The ranking of the level is described in 1/0 format with Yes/No.

COVID-19 has had a impact on the education system. The newly developed system has made a challenge to face in education system. The proper (Darbandi et al. 2022)

Table 10.1 Dataset description.

Attributes	Values	Data type
Gender.	1 = Female 0 = Male	Character
Age	Recorded in years	Integer
Did COVID-19 affect your education	1 = Yes 0 = No	Character
Can you afford the source of online education system.	1 = Yes 0 = No	Character
Did you miss online classes due to lack of electricity or other reasons?	1 = Yes 0 = No	Character
Were lectures conducted according to scheduled time?	1 = Yes 0 = No	Character
Should online education be implemented in a regular education system?	1 = Yes 0 = No	Character
Do you feel lack of communication gap between the student and teacher?	1 = Yes 0 = No	Character
Lack of self-confidence.	1 = Yes 0 = No	Character
Does it affect in clearing doubts.	1 = Yes 0 = No	Character
Did you attend any online course?	0 = Yes 1 = No	Character
How much do you understand during online classes?	0/1/2/3/4/5	Integer
Are you confident about recalling syllabus?	0/1/2/3/4/5	Integer
Do you feel classes to be boring?	0/1/2/3/4/5	Integer
Feeling tired of homework?	0/1/2/3/4/5	Integer
Was it challenging to face online classes?	0/1/2/3/4/5	Integer
Was it easy to manage time during online exams?	0 = Yes 1 = No	Character
Any health issue due to online education?	0 = Yes 1 = No	Character
Any health issues before COVID-19?	0 = Yes 1 = No	Character
Suffered through COVID-19.	0 = Yes 1 = No	Character
Did it cause an impact on your academic schedule.	0 = Yes 1 = No	Character
How much do you understand during offline classes?	0/1/2/3/4/5	Integer
Do you like to attend classes in college?	0/1/2/3/4/5	Integer
Are you confident to attend physical mode classes?	0/1/2/3/4/5	Integer
Are you confident about recalling syllabus?	0/1/2/3/4/5	Integer
Are you feeling tired of attending college?	0/1/2/3/4/5	Integer
Do you feel classes boring?	0/1/2/3/4/5	Integer
Was it challenging to face offline classes?	0/1/2/3/4/5	Integer
Would you like to have MCQ-based exam in the offline exam system?	0 = Yes 1 = No	Character

(Continued)

Table 10.1 (Continued)

Attributes	Values	Data type
Do you feel overloaded with homework in the offline education system?	0 = Yes 1 = No	Character
Lack of confidence with academic performance.	0 = Yes 1 = No	Character
Do you feel conflict while studying college work?	0 = Yes 1 = No	Character
Difficult to pay attention in class.	0 = Yes 1 = No	Character
Do you feel tired of sitting whole day in college?	0 = Yes 1 = No	Character
Do you feel anxiety?	0 = Yes 1 = No	Character
Any health issues post offline classes?	0 = Yes 1 = No	Character
Rank online education system	0/1/2/3/4/5	Integer
Rank offline education system.	0/1/2/3/4/5	Integer

description is analyzed and the system made for the newly updated system is introduced. The survey has been made, and all possible analyses were done with all applied algorithms. The EDA technique was used to make all the graphical analyses, displaying all the analytical ways of representing the analyzed data. The impact of COVID caused a diverse effect on its education system. The online means of the education system was introduced as soon as lockdown was announced, and all colleges, schools and universities, etc., were closed down.

During the lockdown, the academic system provided a newly introduced education system. The introduction of this system increased the complexity for learning. The advantages and disadvantages are described further for analysis. (Alarcón et al. 2021) The big advantage of online education was to learn new technology from any place. Distance learning also came into more existence as the system was created due to closed schools and colleges. The disadvantages of online education so far were many such as it was challenging to attend online classes. The next challenge was in rural areas. Due to lack of knowledge in technology, they were not aware about how to use.

Management of COVID is a measure issue which arises after lockdown. As after reopening everything and making each and every thing happen offline, it was major task to manage and keep social distancing. The symptoms were still active in people even after being cured from COVID. To overcome this situation, good management system (Haris and Al-Maadeed 2021) techniques are described. The methodology is described and substituted with a good technique to follow and maintain a good distancing between people. COVID-19 affected all over the world and caused a drastic change to all fields. The main field which is considered is education. All over the world the students were affected due to this pandemic.

The spread of disease was very fast. The number of deaths was increasing day by day. (Alarcón et al. 2021) The analysis of the symptoms is done to prevent at its early onset. The analysis of the symptom prediction is done. There are three levels of symptom prediction. First one is normal, second is severe, and third one is strongly severe. At the first two stages, one can cure the disease at home. But at the third stage, one must consult the doctor. And this analysis is done manually.

The new system of education during online study made a drastic change in the education of students. The online education system has both advantages as well as disadvantages. The lack of sources to study online education made student behind their peers while having online education. The advantages of distance learning made it easy for students as it can be done any time at any place. (Crick et al. 2021) The research paper gives a conclusion on the impact on the education system of students while studying online as well as offline.

The COVID-19 had been marked as a pandemic by the WHO in 2020. The disease spreads very fast. The impact of symptoms on people is still found today even in those who have been cured from coronavirus. The various symptoms are still existing and can impact people now also after being affected by this disease. Various symptoms are observed which can be cured at home by quarantine or by consulting the doctor when the symptoms are at low level. (Crick et al. 2021) These symptoms cause to people last at least 6 months. The overall COVID-19 situation with its symptoms is described.

The challenges faced due to COVID-19 while attending online classes: The challenges are lack of a source to attend classes, lack of electricity, and related problems. The solution to these challenges is described, and the way of communication is included in it. The distribution of devices expands the reach to students, Vilchez-Sandoval et al. (2021), making the use of devices through online lectures common and familiar, which are included in this research paper.

After COVID-19, new challenges will arrive as students will attend the classes offline. The new measures are to be taken post the COVID-19 impact. The main thing to follow is hygiene and sanitation. Every student must keep social distancing. Every student must wear a mask. (Saxena 2021) Each and every management must be done at school level accurately. As there are many students, care must be taken that no one is affected by another student. Temperature checking and taking precautions that no student with symptoms enters the school premises must be exercised. The new challenging situation is described.

The analysis of various learning methods for the pandemic COVID-19 is done. First, the teacher designs open learning through online mode, divides the students into many groups, gives assignments according (Xiao and Li 2020) to the project implementation in the laboratory, class discussions start with each group, results of student discussions form online portfolios, and further the process is repeated.

The exams held in universities, colleges, and schools were un-proctored. This led to having more chances of cheating and not giving it loyally. The main issue of detecting whether the student has secured marks by their own or by having help by others was a new challenging situation faced by teachers. A brief discussion of challenges faced during the lockdown is provided (Duwi et al. 2020), the exams were to be conducted online, which made the few first days very challenging. The system of online examination led to more confidence in students to score marks as there were many chances to take help as the exams were to be taken at home.

Table 10.2 contains the research papers according to the year 2021–2022 where the techniques and conclusion used are described accordingly. Initially, various

Table 10.2 Comparative studies from year 2020.

Sr. no	Author	Year	Technique used	Conclusion
1	C. Xiao et al.	2020	Decision tree	The epidemic reveals the shortcomings of the current educational situations.
2	O. Yahuarcani et al.	2020	Identification of requirements, content production for the app, content digitization, and application design.	Effects of mobile applications have been analyzed as part of educational services in indigenous communities.
3	T. Supriyatno et al.	2020	Empirical methods (study articles, statistical database, and other publications), methods of theoretical analysis (analysis and synthesis; specification and generalization)	Methods to organize online classes.
4	M. Kamil et al.	2020	Sampling data	Description of the educational institute to prevent social distancing.
5	Daniel SJ et al.	2020	Research and development (R&D)	Post steps to manage study within classroom.
6	Tadesse S et al.	2020	Management survey.	Techniques to manage offline classes.

curfews and different methods were used. But, after seeing the rapid spread of disease, the government decided to impose lockdown. The first lockdown was imposed in March 2020. Each and every place was closed, including schools and colleges. The severe impact on the education system affected students, (Yahuarcani et al. 2020) resulting in students to have their education in the online mode, which led in various health issues too for students. The online mode education system led to students lack confidence, which also affected on campus hiring for their jobs.

Different kinds of medicine were introduced during this period. Steroids were also used for curing. Remdesivir was introduced and given to the patient with strong symptoms of coronavirus. The medical description is made along with the symptoms which were introduced. The analysis of the prediction of the symptoms caused to a person at which level is mentioned. (Supriyatno and Kurniawan 2020) Accordingly, which tablets have to be given at which stage is analyzed along with all its description. The anxiety level in a person increases by sitting at home for a long time. The pressure on mental health was affected largely. Sitting alone and being idle caused a very difficult to face kind of situation. The stress level increased in students. The mental health issues increased. Weight loss and gain become a major issue, which caused a major health issue. (Supriyatno and Kurniawan 2020) Sitting alone and being idle caused more anxiety.

Lack of computer knowledge led students to face more complexity while facing online classes. The students who came from rural areas mainly faced difficulty to attend online classes. The challenges are very difficult to tackle at first sight. But analyzing (Daniel, J., 2020) Education and the COVID-19 pandemic detecting through survey created the possibility of having a good source to communicate and talk with student through online mode.

While attending online classes during lockdown, it was difficult to face offline examinations. The offline examination after offline classes was a big challenge to face for students, which caused stress level increase in students. Offline examination directly started with the theoretical part. (Tadesse and Muluye 2020), which directly also affected many students' marks due to which the annual exam score of the students decreased. The student were in stress of facing offline examination.

The analysis of the education status of every student is taken under review through a Google form. The ranks are given to the questions to give their opinion. The yes/no questions are also included to perform the task and get the opinion of the student. The survey is taken, and only manual analysis is done through this dataset which is collected. The area which the student belongs to has different impacts as in rural areas, there might be more accuracy in lack of having technical knowledge.

The management of the study post COVID-19: After lockdown, the system needs to be run efficiently. The schools and colleges must contain everything

Table 10.3 Comparative studies from year 2021 to 2022.

Sr. no	Author	Year	Technique used	Conclusion
1	V. Bhatt et al.	2022	Perception of e-learning via MyCaptain app	App is made for e-learning during online education.
2	A. Hsu et al.	2022	ICT devices	Computer education.
3	Darbandi et al.	2022	Manual observation.	Management techniques of COVID.
4	R.M. Haris et al.	2021	Survey questions and responses; student survey.	Knowing limitations and suggestions about online education.
5	B. Rawat et al.	2021	Sample, questionnaire, analysis, and qualitative data.	Requirements of source to support post COVID situation.
6	T. Crick et al.	2021	Naïve Bayes	COVID symptom analysis.
7	J. Vilchez-Sandoval et al.	2021	Questionnaire survey, Manual survey.	Solution to challenges between online and offline education system.
8	A. Saxena et al.	2021	Manual survey, KNN	Distance education and its importance.

required to conduct offline classes. The management must be done after reopening of schools and colleges (Table 10.3). The described system of management is mentioned and how to manage the study of students after starting of the offline classes. The main fundamentals to conduct offline classes and to mention the required system to manage the study of the students are mentioned. The system of management post COVID-19 is mentioned.

10.6 Data Visualization and Analysis

Data analysis is carried out to reveal the hidden connections and characteristics that exist in the dataset and enhance the machine learning model's performance. Using EDA (exploratory data analysis), we can analyze the data and visualize the data in graphical format. We will analyze all the columns in the dataset with

"Did COVID-19 affect your education system column" so that we can get insights from that.

- Blue portion indicates that COVID-19 did not affect online education.
- Red portion shows that COVID-19 affected education.
- Level 0–1 contains the LOW level of ranking.
- Level 2–3 contains the MEDIUM level of ranking.
- Level 4–5 contains HIGH level of ranking.

10.6.1 Observations

From Figure 10.2. The analysis of the students' understanding level during online classes is displayed. The graph shows the deep described percent count reviewing,

1) In level [0–1] i.e., Low ranking, 16.67% students are affected by COVID-19 and also, they understand during online classes.
2) In level [2–3] i.e., Medium ranking, 51.33% students are affected by COVID-19 and also, they understand during online classes.
3) In level [4–5] i.e., High ranking, 16.67% students are affected by COVID-19 and also, they understand during online classes.

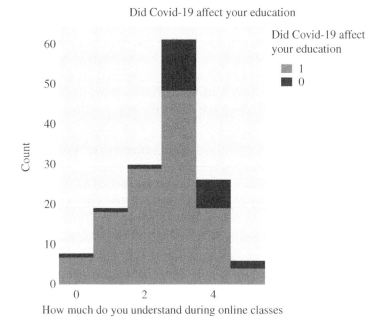

Figure 10.2 Understanding level during online classes.

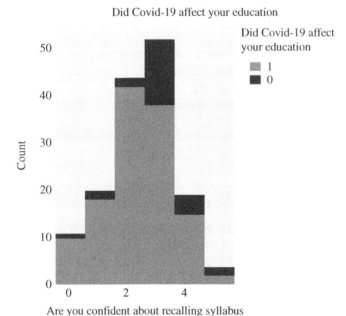

Figure 10.3 Based on recalling teaching contents.

It is observed that 71.77% students reveal that online education does not impact on education, whereas 28.22% students show that lack of understanding due to online education.

From Figure 10.3. The analysis of the students confidence about recalling syllabus in offline classes is displayed. The graph shows the deep described percent count reviewing,

1) In level [0–1] i.e., Low ranking, 18.67% students are affected by COVID-19 and also, they are confident about recalling syllabus in offline classes.
2) In level [2–3] i.e., Medium ranking, 53.33% students are affected by COVID-19 and also, they are confident about recalling syllabus in offline classes.
3) In level [4–5] i.e., High ranking, 11.33% students are affected by COVID-19 and also, they are confident about recalling the syllabus in offline classes.

It is observed that 83.33% students reveal that offline learning helps recalling syllabus, whereas 16.67% students show that recalling learning contents is less in offline learning.

Figure 10.4 displays the analysis of the student's online learning based on Bored feeling. The graph shows the deep described percent count reviewing,

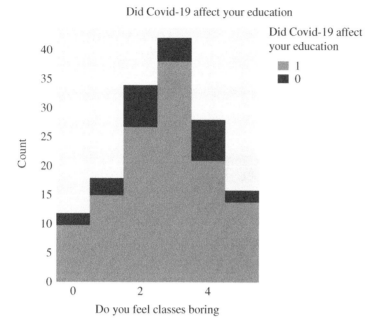

Figure 10.4 Online learning based on bored feelings.

1) In level [0–1] i.e., Low ranking, 16.67% students are affected by COVID-19 and also, they are bored during online classes.
2) In level [2–3] i.e., Medium ranking, 43.33% students are affected by COVID-19 and also, they are bored during online classes.
3) In level [4–5] i.e., High ranking, 23.33% students are affected by COVID-19 and also, they are bored during online classes.

It is observed that 83.33% students reveal that online education feels boring, whereas 16.67% students reveal that online education does not feel bored.

Figure 10.5 shows the analysis of the students offline learning based on tiredness due to writing. The graph shows the deep described percent count reviewing,

1) In level [0–1] i.e., Low ranking, 14% students are affected by COVID-19 and also, they are tired of writing work in offline classes.
2) In level [2–3] i.e., Medium ranking, 45.33% students are affected by COVID-19 and also, they are tired of writing work in offline classes.
3) In level [4–5] i.e., High ranking, 24% students are affected by COVID-19 and also, they are tired of writing work in offline classes.

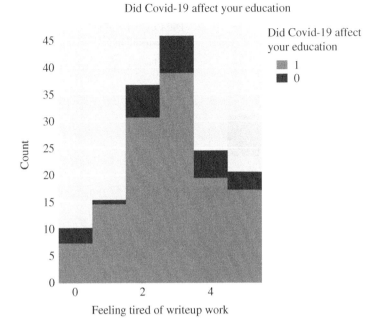

Figure 10.5 Offline learning analysis based on tiredness due to writing.

It is observed that 83.33% students reveal that they are tired due to writing work in offline classes, whereas 16.67% students reveal that they do not feel tired of writing work in offline classes.

Figure 10.6 shows the analysis of the students ranking offline education system. The graph shows the deep described percent count reviewing,

1) In level [0–1] i.e., Low ranking, 6% students are affected by COVID-19 and also, they have given the ranking level for the offline education system according to their choice.
2) In level [2–3] i.e., Medium ranking, 20.67% students are affected by COVID-19 and also, they have given the ranking level for the offline education system according to their choice.
3) In level [4–5] i.e., High ranking, 56.67% students are affected by COVID-19 and also, they have given the ranking level for the offline education system according to their choice.

Figure 10.7 shows the analysis of the students ranking online education system. The graph shows the deep described percent count reviewing,

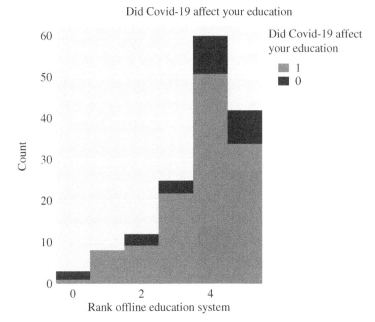

Figure 10.6 Level of ranking in the offline education system.

1) In level [0–1] i.e., Low ranking, 21.33% students are affected by COVID-19 and also, they have given the ranking level for the online education system according to their choice.

2) In level [2–3] i.e., Medium ranking, 48% students are affected by COVID-19 and also, they have given the ranking level for the online education system according to their choice.

3) In level [4–5] i.e., High ranking, 14% students are affected by COVID-19 and also, they have given the ranking level for the online education system according to their choice.

Figure 10.8 shows the analysis of the challenges faced by students in offline classes. The graph shows the deep described percent count reviewing,

1) In level [0–1] i.e., Low ranking, 29.33% students are affected by COVID-19 and also, they have given the ranking according to challenges to face offline classes.

2) In level [2–3] i.e., Medium ranking, 35.33% students are affected by COVID-19 and also, they have given the ranking according to challenges to face offline classes.

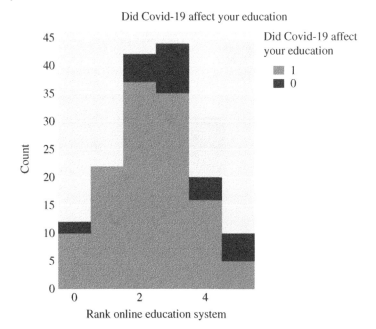

Figure 10.7 Level of ranking of the online education system.

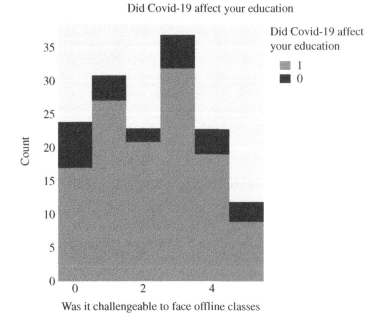

Figure 10.8 Challenges faced in offline classes.

Did Covid-19 affect your education

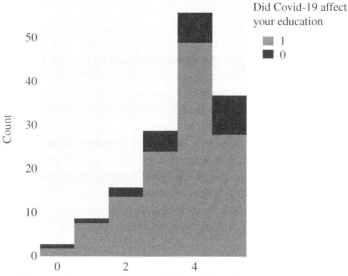

Figure 10.9   Understanding during offline classes.

3) In level [4–5] i.e., High ranking, the 18.67% students are affected by COVID-19 and also, they have given the ranking according to challenges to face offline classes.

Figure 10.9 shows the analysis of what the student feels challengeable to face offline classes. The graph shows the deep described percent count reviewing,

1) In level [0–1] i.e., Low ranking, 29.33% students are affected by COVID-19 and also, they have given the ranking according to challenges to face offline classes.
2) In level [2–3] i.e., Medium ranking, 35.33% students are affected by COVID-19 and also, they have given the ranking according to challenges to face offline classes.
3) In level [4–5] i.e., High ranking, 18.67% students are affected by COVID-19 and also, they have given the ranking according to challenges to face offline classes.

From Figure 10.10, we can say that 83.3% students got affected due to COVID-19 on their education system, whereas 16.7% students think that they did not get affected by COVID-19 on their education system.

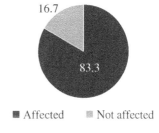

Figure 10.10   Displaying students who got affected on their education due to COVID-19.

10.7 Machine Learning Techniques Implementation

10.7.1 K-Nearest Neighbors

The KNN (K-Nearest Neighbors) is a machine learning algorithm which is used for the classification of problems and the regression of problems. KNN is a very good technique when there is presence of pattern and statistical analysis. The KNN algorithm is primarily based on the feature similarity. The KNN algorithm is a very simple algorithm which gives a high accuracy of data. So, for the analysis, this algorithm is very important. Its k-nearest relatives' method (k-NN) is a pre-regression technique included in regression models for analytical thinking. The input for both scenarios includes the k-test sets that are closest to each other in the training dataset. Whenever KNN is applied for model training, it determines the results of analytical thinking. Giving weight to neighbors' efforts can be helpful for both regression and classification by ensuring that the neighbors who are closer to the average contribute more than those who are farther away.

10.7.2 Decision Tree

Decision Tree is the third algorithm of machine learning techniques which is derived from Artificial Intelligence. Another name of Decision Tree or Decision Tree is also called as support tool of decision. This tool can generate a tree. The Decision Tree algorithm is used for making decisions and giving results of possible outcomes. The Decision Tree algorithm is used for solving decision-related problems. The Decision Tree tracks all paths for giving all possible outcomes. The Decision Tree algorithm is also used for the solving problems of regression and the classification analytics. Numerous rule-based algorithms exist. Among them, J48 was a good one since it builds a strong decision tree using a pruning strategy. Pruning is a technique used to try to get rid of the overly relevant data that overfit the model and makes bad predictions. Finally, a tree is constructed to offer balance, suppleness, and accuracy. By deleting overfitted data, which results in low prediction performance, pruning is a strategy that minimizes the size of the tree. Data are categorized using the DT method recursively until the classification is as accurate as feasible. Best performance on training data is provided by this model. The overall goal is to create a tree that balances accuracy and adaptability.

10.7.3 Random Forest

The random forest algorithm or method is also used for the classification analysis of problems and regression analysis problems. Random forest is the fourth machine learning algorithm which is derived from the artificial intelligence. The random forest algorithm is very flexible. The main role of the random forest algorithm is to find good results and time majority. This algorithm is used for simplicity and variety. The Decision Tree algorithm helps make various multiple decisions with good and accurate prediction. Decision Tree can make lot of trees. This algorithm is used for overfitting problems, which can improve its accuracy.

10.7.4 Support Vector Machine

SVM (Support Vector Machine) is another type of machine learning algorithm. Support Vector Machine can help both the classification and regression analysis problems. SVM is mainly used for the divide classes. Support Vector Machine creates a boundary which divides n-dimensional space into the form of classes. The SVM classifier is the supervised ML technique. SVM is the most widely used categorization approach. It generates a hyperplane that divides two classes. It may generate a hyperplane or series of hyperplanes in three dimensions. This hyperplane can also be suitable for classification and regression. SVM distinguishes instances into specified classes or that can categorize entities that are not even supported at data. Separation is accomplished by using the hyperplane to execute separation between the training dataset point about any class.

10.7.5 Logistic Regression

A statistical modeling approach called logistic regression uses previous observations from the dataset to identify binary outcomes, such as 1/0 (yes/no). With respect to supply of factors, it is applied to anticipate the classified variance. The most widely used identification technology is logistic regression. The main goal of LR is to find the best fit that shows the relation in the target and predictor variable.

10.8 Conclusion

In this research, machine learning algorithms are systematically implemented to draw valuable insight from the data. The suggested approach uses specific classification techniques using Python. In this study, we observe that the Logistic Regression (LR) classifier is better than RF, KNN, DT, and SVM. Basically, we applied the best ML approaches to obtain excellent performance precision. The major goal of this study is to visualize and analyze how much a student is affected by COVID concerning the education system. ML approaches were used to analyze performance of six various classifiers, which was performed effectively.

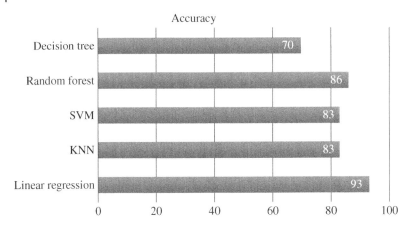

Figure 10.11 Graphical presentation of the accuracy level of the algorithm.

The accuracy for LR (Logistic Regression) is 93%. It is observed that LR classifier performs slightly better than the other classifiers.

References

Alarcón, R. et al. (2021). Correlational analysis of incident factors in the academic performance of university higher education students under the context of COVID-19. *2021 IEEE 1st International Conference on Advanced Learning Technologies on Education &Amp; Research (ICALTER)* [Preprint]. https://doi.org/10.1109/icalter54105.2021.9675086.

Ayuso-Mateos, J.L., Morillo, D., Haro, J.L. et al. (2021). Changes in depression and suicidal ideation under severe lockdown restrictions during the first wave of the COVID-19 pandemic in Spain: a longitudinal study in the general population. *Epidemiology and Psychiatric Sciences* 30: https://doi.org/10.1017/s2045796021000408.

Bhatt, V., Sinha, B.C. and Kumar, V.V.R. (2022). A study on the impact of COVID-19 on students and their perception of e-learning via MyCaptain app. *2022 International Seminar on Application for Technology of Information and Communication (iSemantic)* [Preprint]. https://doi.org/10.1109/isemantic55962.2022.9920441.

Chaturvedi, K., Vishwakarma, D.K., and Singh, N. (2021). COVID-19 and its impact on education, social life and mental health of students: a survey. *Children and Youth Services Review* 121: 105866. https://doi.org/10.1016/j.childyouth.2020.105866.

Crick, T., Knight, C., Watermeyer, R.P. et al. (2021). The international impact of COVID-19 and 'emergency remote teaching' on computer science education practitioners. *Global Engineering Education Conference* [Preprint]. https://doi.org/10.1109/educon46332.2021.9453846.

Daniel, J. (2020). Education and the COVID-19 pandemic. *Prospects* 49 (1–2): 91–96. https://doi.org/10.1007/s11125-020-09464-3.

Darbandi, M., Alrasheedi, A.F., Alnowibet, K.A., et al. (2022). Integration of cloud computing with the Internet of things for the treatment and management of the COVID-19 pandemic. *Information Systems and E-business Management* [Preprint]. https://doi.org/10.1007/s10257-022-00580-5.

Duwi, L.E., Widiyanti, W., and Basuki, B. (2020). Revisiting the impact of project-based learning on online learning in vocational education: analysis of learning in pandemic COVID-19. *IEEE Conference Proceedings* [Preprint]. https://doi.org/10.1109/icovet50258.2020.9230137.

Gournellis, R. and Efstathiou, V. (2021). The impact of the COVID-19 pandemic on the Greek population: suicidal ideation during the first and second lockdown. *Psychiatrikī* https://doi.org/10.22365/jpsych.2021.041.

Haris, R.M. and Al-Maadeed, S. (2021). COVID-19 lockdown-challenges to higher education in qatar. *2021 IEEE 11th IEEE Symposium on Computer Applications &Amp; Industrial Electronics (ISCAIE)* [Preprint]. https://doi.org/10.1109/iscaie51753.2021.9431774.

Hegazi, M.A, Butt, N.S., Sayed, M.H., Zubairi, et al. (2022). Evaluation of the virtual learning environment by school students and their parents in Saudi Arabia during the COVID-19 pandemic after school closure. *PLoS One* 17 (11): e0275397. https://doi.org/10.1371/journal.pone.0275397.

Hsu, A. and Hsu, Y.-F. (2022). Computer Education of the Primary Years Programme Exhibition at International Baccalaureate Schools During the COVID-19 Pandemic. *2022 IEEE 46th Annual Computers, Software, and Applications Conference (COMPSAC)* [Preprint]. https://doi.org/10.1109/compsac54236.2022.00066.

Papadopoulou, A., Efstathiou, V., Yotsidi, V. et al. (2021). Suicidal ideation during COVID-19 lockdown in Greece: prevalence in the community, risk and protective factors. *Psychiatry Research: Neuroimaging* 297: 113713. https://doi.org/10.1016/j.psychres.2021.113713.

Rawat, B., Bist, A.S., and Riza, B.S. et al. (2021). Analysis of examination process during COVID and post COVID In indian context. *2021 9th International Conference on Cyber and IT Service Management (CITSM)* [Preprint]. https://doi.org/10.1109/citsm52892.2021.9588821.

Saxena, A. (2021). Challenges and factors influencing early childhood education in Hong Kong during COVID-19: teachers' perspective. *International Conference on Advanced Learning Technologies* [Preprint]. https://doi.org/10.1109/icalt52272.2021.00040.

Schnell, T. and Krampe, H. (2020). Meaning in life and self-control Buffer stress in times of COVID-19: moderating and mediating effects with regard to mental distress. *Frontiers in Psychiatry* 11: https://doi.org/10.3389/fpsyt.2020.582352.

Supriyatno, T. and Kurniawan, F. (2020). A new pedagogy and online learning system on pandemic COVID 19 era at islamic higher education. *2020 6th International Conference on Education and Technology (ICET)* [Preprint]. https://doi.org/10.1109/icet51153.2020.9276604.

Tadesse, S. and Muluye, W. (2020). The impact of COVID-19 pandemic on education system in developing countries: a review. *Open Journal of Social Sciences* 08 (10): 159–170. https://doi.org/10.4236/jss.2020.810011.

Vilchez-Sandoval, J., Llulluy-Nunez, D., and Lara-Herrera, J. (2021). Work in progress: flipped classroom as a pedagogical model in virtual education in networking courses with the moodle learning management system against COVID 19. *2021 IEEE World Conference on Engineering Education (EDUNINE)* [Preprint]. https://doi.org/10.1109/edunine51952.2021.9429101.

Xiao, C. and Li, Y. (2020). Analysis on the influence of the epidemic on the education in China. *International Conference on Big Data* [Preprint]. https://doi.org/10.1109/icbdie50010.2020.00040.

Yahuarcani, I.O, Saravia Llaja, L.A., Nuñez Satalaya, A.M., Invéntalo, G., Perú, I. et al. (2020). Mobile applications as tools for virtual education in indigenous communities during the COVID-19 pandemic in the peruvian amazon. *2020 3rd International Conference of Inclusive Technology and Education (CONTIE)* [Preprint]. https://doi.org/10.1109/contie51334.2020.00046.

11

Crop Yield and Soil Moisture Prediction Using Machine Learning Algorithms

Debarghya Acharjee, Nibedita Mallik, Dipa Das, Mousumi Aktar, and Parijata Majumdar

Department of Computer Science and Engineering, Techno College of Engineering Agartala, Agartala, Tripura

11.1   Introduction

Crop yield forecasting is one of agriculture's most intricate tasks. It is crucial to decision-making at the international, regional, and local levels and to minimize the overall losses in production. Crop yield may be predicted from soil conditions, agricultural, meteorological, and other factors. Decision support models are widely used to extract crucial agricultural features for crop forecasting (Nyeki and Nemenyi 2022).

It is crucial to predict agricultural production in order to address growing food security problems, particularly in the age of global climate change. In addition to assisting farmers in making wise financial and managerial decisions, accurate yield estimates also help avert famine (Ansarifar et al. 2021).

Big Data technologies and high-performance computers have advanced alongside machine learning, creating new opportunities for measuring and comprehending data-intensive work related to agriculture. Machine learning is defined as the branch of science that enables devices to learn without being explicitly programmed. A growing variety of scientific fields are utilizing machine learning, including robotics, biochemistry, medicine, bioinformatics, aquaculture, economic sciences, and food security (Liakos et al. 2018).

The foundation of the Indian economy is agriculture. The weather significantly affects agricultural productivity in India. Rainfall is mostly necessary for rice cultivation. To assist farmers in maximizing crop production, timely guidance to

Machine Learning Applications: From Computer Vision to Robotics, First Edition.
Edited by Indranath Chatterjee and Sheetal Zalte.

forecast future crop productivity and an analysis must be made. Crop yield prediction is a key problem in agriculture. Based on yield information from previous years, farmers used to forecast their production. Thus, there are various methods or algorithms for this kind of data analytics in crop prediction, and we can predict crop production with the aid of those algorithms (Champaneri et al. 2016).

Algorithms such as deep neural network, artificial neural network, and support vector machine were taken into use in different projects.

Random forest algorithm, K-nearest neighbors, linear regression, and decision tree were used to predict the crop yield, and the best accuracy of 98% was obtained with the KNN algorithm.

Ascertaining the amount of moisture in the soil is essential because water facilitates crop growth by working as a solvent and a carrier of dietary components. The quantity of water that is accessible has an impact on crop output. Soil water is a nutritional element in and of itself. The circulation of water present in the soil regulates soil temperature. Water in the soil is necessary for weathering and soil formation processes. Soil water is required for microorganism metabolic processes. Soil water facilitates chemical and biological activity. Soil is the most important component for plant development, while water is required for photosynthesis (Paul and Singh 2020).

Soil moisture is a key indicator for defining and identifying agricultural drought. In particular, for desert areas, soil moisture evaluation has implementations in identifying initial water deficit situations, advancing drought events for crop yield insecurity and food security conditions, agricultural insurance, and suitable crop planning. Drought badly impacts the developing nations, contributing to social and political instability. As a result, soil moisture modeling and monitoring are becoming increasingly important (Adab et al. 2020).

In recent years, it has become clear that soil fertility and health are closely associated to agricultural sustainability. By maintaining soil fertility, it is also possible to increase present crop yield levels. Water usage can be greatly improved by keeping track of soil moisture level. Because soil moisture is crucial to crop production, its prediction can improve the hydrological processes that are involved in crop growth (Prakash et al. 2018).

Different algorithms like decision tree, support vector machine, and artificial neural network are used in existing literature. In some cases, one method may outperform the other.

For example, the soil moisture estimation results using SVM (support vector machine) modeling are compared to those of the ANN (artificial neural network) and MLR (multilinear regression) models which outperform them (Ahmad et al. 2010).

For soil moisture prediction, we have used naive Bayes regression, linear regression, support vector regression (SVR), and decision tree. Among these algorithms, the decision tree algorithm has shown the best accuracy with a training accuracy of 99% and a testing accuracy of 95%.

11.2 Literature Review

The focus is on predicting the crop yield and monitoring soil moisture depending on the existing data through linear regression, decision tree, KNN, naive Bayes, and SVR. The algorithm utilized here uses the data that are currently available to predict the agricultural yield. The models were created using real data from each state in India, and they were assessed using samples.

In van Klompenburg et al. (2020), the authors have extracted and synthesized the techniques that are being used in agricultural yield prediction research, and they reviewed various literature and obtained 567 pertinent studies from six Internet resources using their search parameters. Approximately 50 studies have been chosen for additional analysis. They looked into these carefully chosen studies, examined the techniques and characteristics employed, and offered recommendations for more research. The investigation shows that temperature, rainfall, and soil type are the three parameters that are most frequently used for soil moisture and crop yield prediction. Artificial neural networks are used in these models.

In (Shahhosseini et al. 2021), the authors have studied and explored the possibility of machine learning and crop modeling to combinedly enhance the estimation of crop production. The major goal is to find which type of model combinations produce the most accurate forecasts, research if a hybrid method between crop modeling and machine learning would improve predictions, and identify the crop modeling elements that are most effectively integrated with ML for corn yield prediction. According to results, ML models' root mean squared error for crop yield can be reduced from 7 to 20% by including modeling of model variables and agricultural production systems simulator as input features.

In (Ansarifar et al. 2021), the authors have introduced a prediction model for crop production prediction that combines the power of machine learning, optimization, and agronomic knowledge having different characteristics. First, it beat cutting-edge machine learning algorithms for yield prediction in three different locations with a relative root mean square error of 8% or less. Furthermore, it discovered a few or more environment-by-management interactions for crop. Third, it enabled agronomists to determine the elements that influence yield in a given site under certain weather management situations. This was accomplished by quantifying agricultural production contributions from weather, soil, and other different characteristics.

In (Kamath et al. 2021), the authors' study offers a quick look into agricultural yield forecasts for a particular area using the random forest algorithm. Crop production forecasting includes a vast quantity of data, making data mining techniques the ideal choice. Data mining is a technique for gathering predicted information from huge databases that had never before been viewed. Making educated judgments is made possible for businesses because of data mining, which aids in the understanding of future patterns and characteristics.

In (Khaki and Wang 2019), the authors created a deep neural network (DNN) technique that took the use of cutting-edge modeling and resolution technologies. The model exhibited a greater degree of prediction accuracy, with an RMSE (root mean square error) of 12%. With optimal climatic data, the RMSE would be 11% of the average return and 46% of the standard deviation. The DNN model was used in decreasing the size of the input space without affecting prediction accuracy much. The computational findings show that this model performs better than other methods employed. It shows that environmental variables had a bigger impact on crop production.

In (Prakash et al. 2018), the authors' study employed machine learning methods to calculate or predict the amount of moisture present in soil over the upcoming days. The methods include SVR, multiple linear regression, and recurrent neural networks. Three separate datasets that were gathered from various web repositories were subjected to these approaches. The mean square error (MSE) and coefficient of determination are used to assess the predictor's performance which obtains best results.

In (Cai et al. 2019), the authors have concluded that by combining the information, studying the time series, and explaining the link among features and predictive variables, meteorological data may provide helpful insights for soil moisture prediction. According to test findings, the deep learning model is practical and useful for forecasting soil moisture. Its data fitting and generalization skills may increase the input characteristics while ensuring high accuracy. In addition, it provides a theoretical framework for water-saving and controlling drought.

In (Paul and Singh 2020), the authors have predicted moisture of the soil for a time period using machine learning approaches such as support vector machine regression, linear regression, principal component analysis, and naive Bayes. These methods were used for distinct crop datasets gathered from different locations gathered for 3 months. The F1-Score was used to assess the predictor's performance.

In (Kanade and Prasad 2021), the authors have created a project that combines IoT (Internet of Things) and machine learning. The hardware is made up of many sensors, a microcontroller, and an IoT module. The sensors will forecast the weather in an area, allowing farmers to use less water. The pH sensor helps detect the pH of the soil and anticipate whether it needs extra water at regular intervals. The main goal is to develop an irrigation system automatically and save water for future use.

In (Vij et al. 2020), the authors developed a model using IoT, that provides a more economically effective and precise solution to meet farming needs. A monitoring system that is designed to efficiently manage problems related to irrigation such as overwatering, soil erosion, and some specific types of crop requires special attention. Because it is commonly known that water is a precious resource, waste of water resources should be minimized. The model will be created by building a distributed wireless sensor network in which each part of the field will be covered by sensor modules to broadcast data on a central server.

11.3 Methodology

The main objective is to predict crop yield in advance using different algorithms. Algorithms were tested for best accuracy and implemented to obtain the output.

Dataset collection: We collected crop data from www.kaggle.com consisting of temperature, humidity, pH value of the soil, rainfall level, nitrogen (N), phosphorus (P), and potassium (K) contents. The data were collected in accordance with all the states in India, and the data about these climatic parameters were gathered on a yearly basis.

Data partitioning: The entire dataset is divided into distinct datasets which include 75% of the data assigned for training the model and 25% data used for testing. For soil moisture monitoring, at first the data were collected using an IoT-based sensor. The collected data were then tested using different algorithms to observe the actual vs predicted value. The algorithm with the best predicted value (decision tree) was chosen to predict the final output.

Data collection from an Arduino sensor board: The sensor board includes different types of sensors, relays, LEDs, and sound buzzers. For data collection, the sensors for soil moisture, humidity, and temperature are used in this project. The sensor board has been expanded by the addition of a Wi-Fi module, which is powered by a micro-USB 2.0 cable. This Wi-Fi extension was required to connect to the data collection REST API interface.

For crop yield prediction, we used the following algorithms for prediction:

1) **Linear regression**: *"It is a simple prediction technique."* *It provides estimates for continuous variables like income and age. A linear connection between one or more independent (y) variables and a dependent (x) variable is demonstrated using the linear regression procedure. It denotes a linear relationship that may be used to assess how the value of the dependent variable changes as a function of the value of the independent variable* (see Figure 11.1). (Maulud and Abdulazeez 2020).

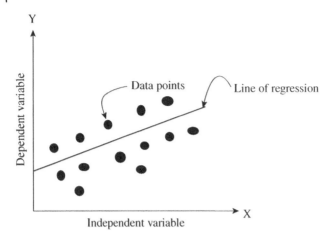

Figure 11.1 Linear regression algorithm. https://www.javatpoint.com/linear-regression-in-machine-learning

2) **Random forest**: *"It is a classification and regression problem method. It constructs decision trees from several samples, using their average for classification and majority vote for regression. It is a classifier that uses a number of decision trees on different subsets of the supplied dataset and takes the average to enhance the dataset's accuracy."* (Abdulkareem and Abdulazeez 2021). A random forest comprises numerous trees, and increasing number of trees will increase the robustness of the prediction model.

3) **KNN algorithm**: *"It is a basic machine learning method that is based on supervised learning and is an abbreviated form of K-Nearest Neighbor. The K-NN method assumes that the new case and the old cases are comparable, and it assigns the new example to the category that is the most similar to the existing categories. K-Nearest Neighbor is a fundamental machine learning approach based on the supervised learning method. The K-NN technique assumes that the new and old examples are comparable and allocates the new example to the category that is closest to the existing categories. After saving all prior data, a new data point is classified using the K-NN algorithm based on similarity."* (see Figure 11.2) (Zhang 2016).

4) **Decision tree**: *"A decision tree, a supervised learning strategy, used to handle classification and regression problems. It is a tree-structured classifier, with internal nodes reflecting the properties of the dataset, branches representing the decision-making process, and each leaf node indicating the classification result. Decision nodes are used to produce choices and have many branches, whereas leaf nodes are the results of decisions and have no extra branches. The features of the submitted dataset are used to carry out the test or make the decision. It is a*

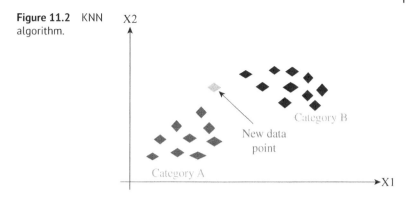

Figure 11.2 KNN algorithm.

graphical representation of all potential responses to a decision or problem based on predetermined circumstances. It is called a decision tree because, like a tree, it starts with the root node and works its way up." (Navada et al. 2011).

For soil moisture prediction, we used the abovementioned algorithms and also used the following:

5) **Support vector machine**: *"Support Vector Machine is a supervised machine learning technique is used for both classification and regression. It finds a suitable hyperplane in an N-dimensional space that unambiguously classifies the input data points. The number of features defines the size of the hyperplane. If there are just two input characteristics datasets, the hyperplane is merely a line. If there are three input characteristics, the hyperplane transforms into a 2-D plane"* (see Figure 11.3) (Wang 2022).

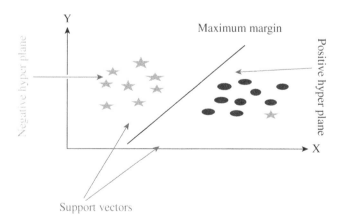

Figure 11.3 Support vector machine algorithm. https://www.javatpoint.com/machine-learning-support-vector-machine-algorithm

6) **Naive Bayes regression**: *"It is a supervised learning approach based on the principle of Bayes theorem. It is most commonly used in text categorization having a large training dataset. It contributes to the creation of fast machine learning models capable of generating accurate predictions. It assumes there is strong independence among the features. It provides predictions based on the likelihood of an object that will occur since it is a probabilistic classifier. Naive Bayes algorithms are commonly employed in spam filtration, sentiment analysis, and article categorization"* (Frank et al. 2000).

11.4   Result and Discussion

Here, the accuracy table (see Table 11.1) generated for crop yield prediction using different machine learning algorithms is shown. The KNN shows the highest accuracy of 98.18% along with a standard deviation of 0.66%.

Table 11.2 shows the training and testing accuracy of the algorithms used for soil moisture monitoring. Here, the decision tree algorithm shows the best accuracy.

Table 11.1 Accuracy and standard deviation table for crop yield prediction.

Algorithm	Accuracy	Standard deviation
Random forest	96.67	0.691015
K-nearest neighbor	98.18	0.661450
Decision tree	92.43	2.520343
Linear regression	93.15	0.656250
Support vector regression	94.68	0.672050
Naive Bayes classifier	95.23	2.512243

Table 11.2 Accuracy table for soil moisture monitoring.

Algorithms	Training accuracy	Test accuracy
Linear regression	0.2237392474704788	0.24648312674488826
Support vector regression	0.4170893044090419	0.48205922182776655
Naive Bayes classifier	0.555555555555556	0.95267448882612222
Decision tree	0.999987777799999	0.95555555555569999
K-Nearest neighbor	0.3222222222222224	0.35999999122249
Random forest	0.9862287777799999	0.94655555555569

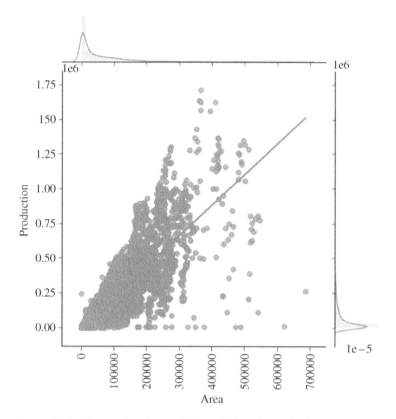

Figure 11.4 Scatter plot of crop yield prediction obtained using linear regression.

The following output shows the production of rice in the state of Tripura (see Figure 11.4). Production per unit area decreases with increasing area.

For soil moisture prediction, Table 11.3 shows the resultant data of actual values compared to the predicted values of soil moisture data. The output was obtained using the linear regression model. The table depicts the moisture of the soil, where the actual value is obtained using an IoT-based sensor while the predicted value is obtained using a linear regression algorithm.

11.5 Conclusion

The improvements in accuracy of yield prediction have been observed using different machine learning algorithms. Water use can be greatly improved by keeping track of soil moisture because soil moisture is crucial to crop production.

Table 11.3 Actual vs predicted value in soil moisture prediction.

Sl. no	Actual	Predicted
0	0.6999	0.506906
1	0.4751	0.526563
2	0.7400	0.632289
3	0.6276	0.548861
4	0.5210	0.531265
5	0.7312	0.603859
6	0.5777	0.647093
7	0.4633	0.531514
8	0.4604	0.504015
9	0.4741	0.512689
10	0.4907	0.516610
11	0.7058	0.518670
12	0.7361	0.640131
13	0.5601	0.583322
14	0.4673	0.506026
15	0.7204	0.600918
16	0.5445	0.531315
17	0.4790	0.525483
18	0.5601	0.654886

It improves the process involved in crop growth. The regression models of K-nearest neighbor used for soil moisture and decision tree for crop yield prediction show better results than other state-of-the-art machine learning algorithms where datasets are acquired for the state of Tripura, and they identify many environment factors responsible for crop yield and soil moisture monitoring. The variables that favorably or negatively affect the crop yield are identified as they showed the contributions of soil and weather management, and their relationship to crop yield. Environmental data were collected using real-time sensors for a period of 3 months in the year 2020. After the identification of extensive related literature analysis that applied several machine learning algorithms for prediction, we extracted and synthesized the applied algorithms for prediction. Thus, machine learning algorithms are suitable for crop yield prediction. This type of model can be applied to various areas that can help obtain outstanding results for predicting the yield and quality of crops, provided insights to farmers, and offer predictions regarding livestock production. We hope that this research helps pave the way for further research for the development of more accurate hybrid machine learning algorithms for crop yield and soil moisture prediction.

References

Abdulkareem, N.M. and Abdulazeez, A.M. (2021). Machine learning classification based on radom forest algorithm: a review. *International Journal of Science and Business* 5 (2): 128–142.

Adab, H., Morbidelli, R., Saltalippi, C. et al. (2020). Machine learning to estimate surface soil moisture from remote sensing data. *Water* 12 (11): 3223. https://doi.org/10.3390/w12113223.

Ahmad, S., Kalra, A., and Stephen, H. (2010). Estimating soil moisture using remote sensing data: a machine learning approach. *Advances in Water Resources* 33 (1): 69–80. https://doi.org/10.1016/j.advwatres.2009.10.008.

Ansarifar, J., Wang, L., and Archontoulis, S.V. (2021). An interaction regression model for crop yield prediction. *Scientific Reports* 11 (1): 1–14. https://doi.org/10.1038/s41598-021-97221-7.

Cai, Y., Zheng, W., Zhang, X. et al. (2019). Research on soil moisture prediction model based on deep learning. *PloS One* 14 (4): e0214508. https://doi.org/10.1371/journal.pone.0214508.

Champaneri, M., Chachpara, D., Chandvidkar, C., and Rathod, M. (2016). Crop yield prediction using machine learning. *Technology* 9: 38.

Frank, E., Trigg, L., Holmes, G., and Witten, I.H. (2000). Naive Bayes for regression. *Machine Learning* 41 (1): 5–25. https://doi.org/10.1023/a:1007670802811.

Kamath, P., Patil, P., Shrilatha, S., and Sowmya, S. (2021). Crop yield forecasting using data mining. *Global Transitions Proceedings* 2 (2): 402–407. https://doi.org/10.1016/j.gltp.2021.08.008.

Kanade, P. and Prasad, J.P. (2021). Arduino based machine learning and IOT smart irrigation system. *International Journal of Soft Computing and Engineering (IJSCE)* 10 (4): 1–5. https://doi.org/10.35940/ijsce.d3481.0310421.

Khaki, S. and Wang, L. (2019). Crop yield prediction using deep neural networks. *Frontiers in Plant Science* 10: 621. https://doi.org/10.3389/fpls.2019.00621.

Liakos, K.G., Busato, P., Moshou, D. et al. (2018). Machine learning in agriculture: a review. *Sensors* 18 (8): 2674. https://doi.org/10.3390/s18082674.

Maulud, D. and Abdulazeez, A.M. (2020). A review on linear regression comprehensive in machine learning. *Journal of Applied Science and Technology Trends* 1 (4): 140–147. https://doi.org/10.38094/jastt1457.

Navada, A., Ansari, A.N., Patil, S., and Sonkamble, B.A. (2011). Overview of use of decision tree algorithms in machine learning. *2011 IEEE Control and System Graduate Research Colloquium*, Shah Alam, Malaysia (27–28 June 2011), pp.37–42. IEEE. https://doi.org/10.1109/icsgrc.2011.5991826.

Nyeki, A. and Nemenyi, M. (2022). Crop yield prediction in precision agriculture. *Agronomy* 12 (10): 2460. https://doi.org/10.3390/agronomy12102460.

Paul, S. and Singh, S. (2020). Soil moisture prediction using machine learning techniques. *2020 The 3rd International Conference on Computational Intelligence and Intelligent Systems* (13–15 November 2020), pp. 1–7. ACM. https://doi.org/10.1145/3440840.3440854.

Prakash, S., Sharma, A., and Sahu, S.S. (2018). Soil moisture prediction using machine learning. *2018 Second International Conference on Inventive Communication and Computational Technologies (ICICCT)* (20–21 April 2018), pp.1–6. IEEE. https://doi.org/10.1109/icicct.2018.8473260.

Shahhosseini, M., Hu, G., Huber, I., and Archontoulis, S.V. (2021). Coupling machine learning and crop modeling improves crop yield prediction in the US corn belt. *Scientific Reports* 11 (1): 1–15. https://doi.org/10.1038/s41598-020-80820-1.

Van Klompenburg, T., Kassahun, A., and Catal, C. (2020). Crop yield prediction using machine learning: a systematic literature review. *Computers and Electronics in Agriculture* 177: 105709. https://doi.org/10.1016/j.compag.2020.105709.

Vij, A., Vijendra, S., Jain, A. et al. (2020). IoT and machine learning approaches for automation of farm irrigation system. *Procedia Computer Science* 167: 1250–1257. https://doi.org/10.1016/j.procs.2020.03.440.

Wang, Q. (2022). Support vector machine algorithm in machine learning. *2022 IEEE International Conference on Artificial Intelligence and Computer Applications (ICAICA)* (24–26 June 2022), pp. 750–756. IEEE. https://doi.org/10.1109/icaica 54878.2022.9844516.

Zhang, Z. (2016). Introduction to machine learning: k-nearest neighbors. *Annals of Translational Medicine* 4 (11): https://doi.org/10.21037/atm.2016.03.37.

12

Multirate Signal Processing in WSN for Channel Capacity and Energy Efficiency Using Machine Learning

Prashant R. Dike[1], T. S. Vishwanath[2], V. M. Rohokale[3], and D. S. Mantri[4]

[1] Research Scholar, Bheemanna Khandre Institute of Technology, Karnataka, India
[2] Bheemanna Khandre Institute of Technology, Karnataka, India
[3] Sinhgad Institute of Technology and Sciences, Maharashtra, India
[4] Sinhgad Institute of Technology, Maharashtra, India

12.1   Introduction

The wireless sensor network (WSN) is the fastest growing field in the era of communication. The parameters that are to be focused are throughput and delay and energy consumption. Increasing throughput of the system enhances the performance of the system. In the WSN, information or data assembled for the data is sensed and further processed. But, these sensors in the WSN are limited to their performance in terms of bandwidth. New Machine Learning technology approaches are used to overcome such limitations. This chapter addresses the issue related to throughput and provides an improved algorithm for enhancement of parameters. In a multichannel network, the selection of link rate is difficult. In this paper, a new network coding in the WSN is proposed with the intention of maximizing the throughput of the network. Here, the improved meta-heuristic algorithm called modified Lion Algorithm (LA) will be used to solve the problem of network coding in the WSN.

WSNs enjoy great benefits due to their low-cost, small-scale factor, and smart sensor nodes. Not only can they be employed in cumbersome and dangerous areas of interest but also for monitoring or controlling the regions. Recently, the WSN has attracted much interest to develop for various applications such as environmental monitoring, military reconnaissance, target tracking, health,

Machine Learning Applications: From Computer Vision to Robotics, First Edition.
Edited by Indranath Chatterjee and Sheetal Zalte.
© 2024 The Institute of Electrical and Electronics Engineers, Inc.
Published 2024 by John Wiley & Sons, Inc.

surveillance, and robotic exploration (Akyildiz et al. 2002). A WSN consists of low powered sensors, which have sensing, processing, and communication capabilities (Vieria et al. 2013).

Improving energy efficiency and decreasing energy consumption are still a main challenge of WSNs (Khalily-Dermany et al. 2017). Energy-efficient techniques, such as network coding, have received extensive attention from industrial and academic societies to decrease energy consumption and improve network performance (Karl and Willig 2007). Network coding is a technique in which different mathematical operations are used to combine different nodes, which reduces the number of transmitted packets. Network coding was first proposed for wired networks to solve the bottleneck problem and to increase the throughput (Senouci et al. 2012). Network coding (NC) gives the required change in the wireless network by using broadcast nature of WSN channels Moreover, it tends to reduce power consumption in comparison with the traditional simple store-and forward approach (Katti et al. 2008).

A network which uses coding techniques permits different nodes to encode different incoming packets instead of just forwarding them. This powerful theory can improve the network load and enhance network robustness by employing path diversity. So, in addition to prolonging the lifetime of WSNs, NC improves data security. As a result, NC, for exploiting the correlations of sensor readings in WSNs, has become an attractive topic (Rout and Ghosh 2014).

The majority of the NC schemes in use nowadays are based on algebraic theory (Tonneau et al. 2015). Whereas earlier schemes, such as the traditional XOR-coding scheme and the Deterministic Linear Network Coding scheme, were deterministic in nature, and the more common schemes in use today are non-deterministic, meaning they are free from the constraint of having packet feedback information for every transmitted packet from all the receivers (Sun et al. 2014). In this section, we shall take a look at some common coding schemes, namely Random Linear Network Coding (RLNC), Triangular Network Coding (TNC), and Opportunistic Network Coding (ONC). Physical-layer network coding as a concept was proposed in 2006 for application in wireless networks. The main idea of PNC is to take advantage of the natural network coding operation that occurs when electromagnetic (EM) waves are superimposed on one another (Faraji-Dana and Mitran 2013). This simple idea leads to numerous advancements (Yang et al. 2010), with subsequent works by various researchers leading to many new results in the domains of wireless communication, wireless information theory, and wireless networking (Yang et al. 2015). With this in mind, PNC is the future of NC in wireless networks, and thus it is important we understand it. Further, research studies are still going on in the field of implementing network coding in WSNs (Mirjalili et al. 2014) for the betterment and new innovation. Combining topology control and network coding optimizes the lifetime in wireless-sensor networks (Khalily-Dermanya et al. 2019).

12.2 Energy Management in WSN

Wireless sensor networks (WSNs) can be defined as the networks in which individual nodes, known as sensors, collect and send data collected from their surrounding environment to a central hub, where the data are processed and the results are transmitted over the Internet to far-flung users (Yin et al. 2016). Power management of the sensor node's numerous components, shown in Figure and discussed in (Lee et al. 2009), is crucial to ensuring the availability of communication between the two points: sensor node and base station at all times (Xu et al. 2016). The communication subsystem of the sensor node, primarily controlled by the radio, is the dominant source of energy consumption during transmission, as illustrated in Figure 12.1, despite the installation of these energy management techniques.

12.3 Different Strategies to Increase Energy Efficiency

Some strategies may be used by the routing protocol to increase energy efficiency and network longevity. Several power managing approaches are listed as below:

- **Power models**: The power model of the sensor devices may help improve throughput in any routing algorithm (Tonneau et al. 2015). A properly defined energy model may give a clearer calculation of each network node residual power. It streamlines and focuses monitoring. A model with a comprehensive viewpoint and an appropriate approach may be able to prolong the network's lifetime (Borges et al. 2014).
- **Reduce collisions**: Information must approach the BS without being interfered with via a routing mechanism. In a crowded environment, the protocol

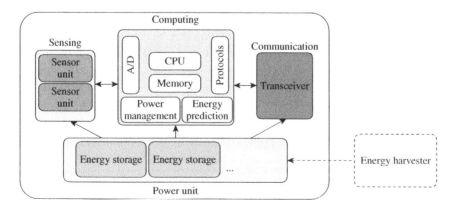

Figure 12.1 Wireless sensor node.

must guarantee that each node interacts. Otherwise, data re-transmission may occur, which has a direct effect on the program's efficiency.

- **Reduce data packets workload**: The sensor consumes much more power while information transfer. Routing methods for the neighbor's information need a high set of control messages to be sent between sensors for path detection and maintaining. The directing must minimize the unnecessary transmission of controls packets in the scheme. The duration of the control packets can influence total power usage (Kennedy et al. 1995).

- **Allow for multi-hop interactions**: Straight information transfer always uses a lot of power than multi-hop interactions. When directly interacting, the sensor must boost radio signal power, increasing energy usage at each node. The routing must handle these issues in addition to increasing the efficiency of energy (Boothalingam et al. 2018).

- **Making use of power aware MAC protocols**: The sensor examines its environment, provides information, and transmits it to the sinks (Chen et al. 2016). Sensors must rest if they are not discovering or transmitting. As a consequence, for system efficiency improvements, a suitable MAC protocol is required.

- **Provisioning**: Remaining power information is important in a distributed network where each node must maintain itself. Using the power model, every node calculates its residual power. The routing must balance the load among sensors such that more work is allocated to nodes with a lot of residual power and less work is given to nodes with low residual energy. The use of an effective load balancing strategy improves energy efficiency.

- **Adjustment of transmission range**: A wireless communication is a multi-hop system in which information must be transmitted via intermediate nodes to reach its target. During delivery, it is often found that the next available relay nodes are always near the sender node. As a consequence, rather than transmitting data at maximum power, the RSSI may be used to limit transmission strength. This approach may help reduce energy use while simultaneously improving network lifespan.

- **Data aggregation**: Relevant packets of data may be combined at some point, and the data collected may be transmitted to the sink. The technique of clustering similar data together lowers network bandwidth. Reduced traffic means fewer accidents and lower energy use. To extend the network's lifespan, the aggregation method must be incorporated in the routing algorithm.

12.4 Algorithm Development

NC and change in the rate are the important parameters for enhancement of throughput and delay. Combining NC and diverse rate increases potential gain. Wireless topology with number of nodes is as shown in Figure 12.2. The number of nodes considered is 5. The path between node 1 and 2 is 1 Mbps. The path between the other points is 11 Mbps. Both the positions 1 and 2 transmit the information A and B, respectively. In a wireless transmission, the links are not independent. Node 1 also transmits data to node 3 with the same rate. Similarly, node 1 sends data to node 4 with the same rate. The size of data is assumed to be 11 Mbps for all the nodes.

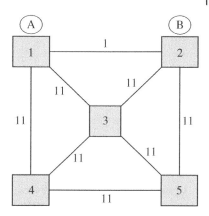

Figure 12.2 A multi-rate network example.

For transmitting information from A and B from position 1 and 2, there are three different ways. These techniques shows that NC and diverse rate in combination enhances throughput and delay. It also increases energy efficiency with the combination of change in rate and network coding.

First, consider data sending with diverse rate unaware of NC

Second, technique data sending with change in rate with no NC

Third, technique data sending with rate diversity aware of NC.

When transmitting information from position 1 to 4 for position 2, it is not possible to transmit the information to 3 as position 3 is on the same range from 1 and 4. It is shown in Figure 12.2.

The time required for sending node 1 data to nodes 2, 3, and 4 is approximately 11 unit. The time required for sending node 2 data to nodes 1, 3, and 5 is approximately 11 unit. Position 3 transmitting information B to position 4 and time required is 1 unit.

Node no 3 data packet A to node 5 and time required is 1 time unit.

The time required to complete the operation is 24 time units.

NC and rate diversity enhances time and throughput. Node 3 transmits A XOR B to nodes 4 and 5 in a single transmission. Node 4 has data related to A and can get data packet B from XOR operation. Node 5 has data related to B and can get data packet A from XOR operation. The conditional operations for three cases data sending without NC, data sending with change of rate with no network coding, and data sending with network coding is shown in Figures 12.3, 12.4 and 12.5 respectively.

Different nodes have different ways to adopt their rates. The total time required for sending the data packet is only 4 units

Figure 12.3 Data sending without NC.

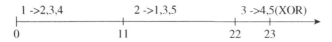

Figure 12.4 Data sending with change of rate with no network coding.

Figure 12.5 Data sending with network coding.

Combination of NC and diverse ate enhances the time by 3 units. NC time decreases to 3 units as added NC and rate diversity.

The information packet which is sent from node 1 to A with the rate of 11 Mbps.

The information packet from node 2 is sent to B to nodes 3 and 5 with rate of 11 Mbps.

Then, node 3 sends (at 11 Mbps) the XOR message to nodes 4 and 5, and also to nodes 1 and 2.

With XOR operation, node 1 will get information of B.

The explanation is given in Table 12.1

In communication, usually power consumption plays an important role, so there are some ways such as control on topology used and NC to decrease the activity of sensor transceivers. If such techniques are utilized simultaneously, then the overall performance does increase as expected. In the WSN, the linear NC has been shown to improve the performance of network throughput and reduce delay. However, the NC condition of existing NC aware routings may experience the problem of false-coding effect in some scenarios and usually neglect node energy, which greatly influences the energy efficiency performance. Hence, this paper proposes a new NC in WSN with the intention of maximizing the throughput of the network. Here, the improved meta-heuristic algorithm called Improved Lion Algorithm (LA) will be used to solve the problem of NC in WSN. The main intention of implementing improved optimization is to maximize the throughput of the network from the source to destination node. While solving the current optimization problem, it has considered few constraints like time share constraint, data-flow

Table 12.1 Comparison of network coding.

Node transformation	Change of rate and no network coding	Change of rate with network coding	Rate diversity-aware with network coding
From 1 to 2,3,4	Total time – 24 unit	Total time – 23 unit	Total time – 3 unit (X-OR)
From 2 to 1,3,5			
From 3 to 4			
From 4 to 5			

Algorithm 12.1

1) Set the values of Y^{male}, Y^{female}, Y^{nomad}
2) Find values of $h(Y^{male})$, $h(Y^{female})$, $h(Y^{nomad})$
3) Initialize $h^{ref} = h^{male}$ and $g = 0$
4) Reserve the values of y^{male} and $h(y^{male})$
5) The fertility evaluation is conducted
6) The cub pool is obtained by performing the crossover and mutation.
7) Find the mutation rate
8) Find the values of coefficient vector J
9) The new cub y^{new}_{cub} is computed
10) Y^{male_cub} and Y^{female_cub} is obtained by performing the gender clustering
11) Set the value of $D_{cub} = 0$
12) The cub growth task is executed
13) The territorial defence is done; if the results obtained from defence is 0 then update the values of y^{male} and $h(y^{male})$
14) If $D_{cub} < D_{max}$, the cub growth is executed
15) The territorial defence is carried out on the values of updated Y^{male}, Y^{female} is obtained.
16) The value of g is increased by 1.
17) If the condition of termination is not met then once again update the values
18) Update the values of y^{male} and $h(y^{male})$. Continue the evaluation process.

---

constraint, and domain constraint. Thus, the multi-rate link layer broadcasts and NC can be mutually combined with new optimization algorithms for enhancing the network throughput. The result of comparison is shown in Figure 12.6 and Table 12.2 gives details of Performance using proposed Algorithm 12.1.

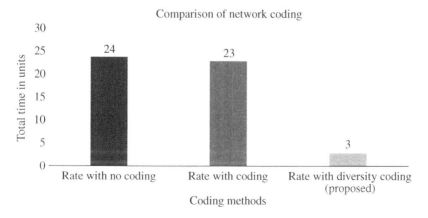

Figure 12.6 Comparison of time.

Table 12.2 Performance table using the proposed algorithm.

Sr.no	Compared with algorithm	Better performance in %
1	FF	48.00
2	PSO	55.17
3	GWO	38.09
4	LA	1

12.5 Results

Proposed work was carried out using MATLAB, and results are compared with some of the existing systems. The simulation area was taken as 100×100 m. Different node configurations are taken for calculating the throughput of the system using the proposed algorithm. The values of energy minimization are compared with those of the proposed system algorithm. The number of nodes used for performance comparison is 20. It is proved from the proposed algorithm that at 500 iterations, we got best result from the improved lion algorithm as shown in Figure 12.7.

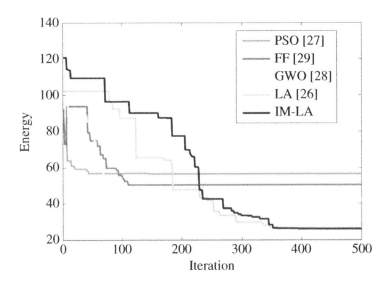

Figure 12.7 Comparison of energy minimization.

The performance of the proposed algorithm in Figure 12.7 is compared with that of existing partial swarm optimization (PSO), Firefly (FF), Gray wolf optimization (GWO), and lion algorithm (LA). Table 12.2 shows the detail performance in percentages when proposed algorithm is used.

12.6 Summary

NC with multi-rate data system increases the throughput of system, which was proven by using the proposed algorithm. It was also proven that for at 500 iterations, the results found were best compared to those of the existing system. The performance of the proposed algorithm is compared with that of existing partial swarm optimization (PSO), Firefly (FF), Grey wolf optimization (GWO), and lion algorithm (LA). The main purpose of the proposed algorithm was to minimize delay and increase in the throughput of the system. Results taken from this paper are greatly useful for future enhancement in NC and change in rate of coding. In this paper, we have proved that combining NC and change in the rate can increase throughput of the network in different applications. Also, machine learning helps analyzing the data and improving network performance.

References

Akyildiz, I.F., Su, W., Sankarasubramaniam, Y. et al. (2002). A survey on sensor networks. *IEEE Communications Magazine* 40 (8): 102–114.

Boothalingam, R. (2018). Optimization using lion algorithm: a biological inspiration from lion's social behavior. *Evolutionary Intelligence* 11 (1–2): 31–52, 2018.

Borges, L., Velez, F., and Lebres, A. (2014). Survey on the characterization and classification of wireless sensor networks applications. *IEEE Communication Surveys and Tutorials* 16 (4) (2014): 1860–1890. https://doi.org/10.1109/COMST.2014.2320073.

Chen, S., Zhao, C., Wu, M. et al. (2016). Compressive network coding for wireless sensor networks: spatio-temporal coding and optimization design. *Computer Networks* 108: 345–356.

Faraji-Dana, Z. and Mitran, P. (2013). On non-binary constellations for channel-coded physical-layer network coding. *IEEE Transactions on Wireless Communications* 12 (1): 312–319, 2013.

Karl, H. and Willig, A. (2007). *Protocols and Architectures for Wireless Sensor Networks*. Wiley https://doi.org/10.1002/0470095121.

Katti, S., Rahul, H., Hu, W. et al. (2008). Xors in the air: practical wireless network coding. *IEEE/ACM Transactions on Networking (ToN)* 16 (3): 497–510. https://doi.org/10.1109/tnet.2008.923722.

Kennedy J and Eberhart, R. (1995). Particle swarm optimization. *Proceedings of IEEE international conference on neural networks, IV*, Perth, WA (27 November 1995–1 December 1995), pp 1942–1948. https://doi.org/10.1109/ICNN.1995.488968.

Khalily-Dermany, M., Shamsi, M., and Nadjafi-Arani, M.J. (2017). A convex optimization model for topology control in network-coding-based-wireless-sensor networks. *Ad Hoc Networks* 59: 1–11.

Khalily-Dermanya, M., Nadjafi-Arani, M.J., and Doostali, S. (2019). Combining topology control and network coding to optimize lifetime in wireless-sensor networks. *Computer Networks* 162: 24.

Lee, J.-Y., Kim, W.-J., Baek, J.-Y., and Suh, Y.-J. (2009). A wireless network coding scheme with forward error correction code in wireless mesh networks. *Proceedings of the Globecom'09*, Honolulu, HI (30 November 2009–4 December 2009), pp. 1–6. IEEE. https://doi.org/10.1109/GLOCOM.2009.5426042.

Mirjalili, S.M., Mirjalili, A., and Lewis, A. (2014). Grey wolf optimizer. *Advances in Engineering Software* 69: 46–61.

Rout, R.R. and Ghosh, S.K. (2014). Adaptive data aggregation and energy efficiency using network coding in a clustered wireless sensor network: an analytical approach. *Computer Communications* 40: 65–75.

Senouci, M.R., Mellouk, A., Senouci, H., and Aissani, A. (2012). Performance evaluation of network lifetime spatial-temporal distribution for WSN routing protocols. *Journal of Network and Computer Applications* 35 (4): 1317–1328.

Sun, B., Gui, C., Song, Y., and Chen, H. (2014). A novel network coding and multi-path routing approach for wireless sensor network. *Wireless Personal Communications* 77 (1): 87–99.

Tonneau, A.-S., Mitton, N., and Vandaele, J. (2015). How to choose an experimentation platform for wireless sensor networks? A survey on static and mobile wireless sensor network experimentation facilities. *Ad Hoc Networks* 30 (7) (2015): 115–127. https://doi.org/10.1016/j.adhoc.2015.03.002.

Vieira, L.F.M., Gerla, M., and Misra, A. (2013). Fundamental limits on end-to-end throughput of network coding in multi-rate and multicast wireless networks. *Computer Networks* 57 (17): 3267–3275.

Xu, D., Sun, X., Cheng, H., and Wu, X. (2016). Cluster-level based link redundancy with network coding in duty cycled relay wireless sensor networks. *Computer Networks* 99: 15–36.

Yang, X.-S. (2010). *Nature Inspired Metaheuristic Algorithms*, 2e. London: Luniver Press.

Yang, L., Yang, T., Yuan, J., and An, J. (2015). Achieving the near-capacity of two-way relay channels with modulation-coded physical-layer network coding. *IEEE Transactions on Wireless Communications* 14 (9): 5225–5239, 2015.

Yin, J., Yang, Y., Wang, L., and Yan, X. (2016). A reliable data transmission scheme based on compressed sensing and network coding for multi-hop-relay wireless sensor networks. *Computers and Electrical Engineering* 56: 366–384.

13

Introduction to Mechanical Design of AI-Based Robotic System

Mohammad Zubair

Department of Mechanical Engineering, School of Engineering and Technology, Adamas University, Kolkata, West Bengal, India

13.1 Introduction

The combination of robotics and AI has the potential to enable robots to perform increasingly complex and autonomous tasks by leveraging machine learning and other AI techniques to improve their ability to sense, reason, and act in the world. The goal is to create robots that can learn from their experiences, adapt to changing situations, and make decisions on their own without human intervention. This level of autonomy is measured by the ability to foresee the future and anticipate the consequences of actions, which is a key aspect of human intelligence. While significant progress has been made in developing specialized autonomous robots for specific tasks, such as driving a car, flying a drone, or picking up objects, creating robots that exhibit human-like intelligence and cognitive abilities remains a challenging and elusive goal. There are many technical and scientific challenges that need to be overcome to achieve this level of artificial intelligence, such as developing more advanced machine learning algorithms, improving the ability of robots to perceive and interpret the world, and creating more robust and adaptable control systems. Robots that can perform specialized autonomous tasks like driving a car are a type of autonomous vehicle (Hudson et al. 2019), flying in both natural and artificial environments (Floreano and Wood 2015), swimming (Fukuda et al. 1994), carrying boxes and materials over various terrains, typically referred to as mobile robots or autonomous mobile robots (Wang et al. 2020), and picking up objects, typically referred to as robotic arms or manipulators. These

Machine Learning Applications: From Computer Vision to Robotics, First Edition.
Edited by Indranath Chatterjee and Sheetal Zalte.
© 2024 The Institute of Electrical and Electronics Engineers, Inc.
Published 2024 by John Wiley & Sons, Inc.

robots are designed to manipulate objects with a high degree of precision and accuracy, and they are often used in manufacturing and industrial settings to perform tasks such as assembly, welding, and packaging (Singh et al. 2013; Javaid et al. 2021). However, developing a system that exhibits human-like intelligence remains elusive. Despite these challenges, the potential benefits of creating intelligent robots are enormous, ranging from improved safety and efficiency in manufacturing and logistics to new forms of human–robot collaboration in healthcare, education, and other domains.

For the safe performance of the robot, calibration of the robot is the most desirable task before the robot can be deployed for doing tasks. One such calibration method has been discussed in (Mukherjee et al. 2013a), where a particular open-chain serial manipulator can be converted to a closed chain in order to calibrate a wearable exoskeleton (an open-loop model) deployed for teleoperation of the nuclear pellet in a nuclear power plant. To control a KUKA KR5 robot that is not easily accessible, teleoperation is carried out using an upper limb exoskeleton as the master. The task's design and implementation concerns have been examined (Mukherjee et al. 2013b).

An efficient method to utilize the maximum capability of the actuator is to gravity-compensate the robot using a mechanical system. For the gravity compensation of a robot manipulator that resembles a human arm, a parallelogram mechanism is developed (Zubair and Jung 2021). To account for the system's self-weight, a gravity compensation system is created. The parallelogram design is supported by experimental tests of gravity compensation.

A dynamic model is developed for robots in order to control the system efficiently. A work by (Zubair et al. 2022) discusses a mathematical model of an arm that is necessary to analyze the dynamics of hand tremors. An arm is thought of as extending from the shoulder to the hand, and it normally consists of three joints: the shoulder, elbow, and wrist. To create a mathematical model of the human arm, numerous scholars have carried out research (Nagarsheth et al. 2008). In order to create an orthosis to support the human upper arm, the kinematics of the human arm were investigated using a position sensor (Ramanathan et al. 2000).

Serial manipulators with large kinematic workspaces, like the KUKA KR125, have been utilized to manipulate the CVJ in the clinical literature (Helgeson et al. 2011). Using manipulators in a parallel configuration allows for greater force-moment loading to be applied for the same size actuator for large rotations in a small Euclidean work volume (Tsai 1999).

In this chapter, a serial (open chain) and parallel (closed chain) manipulator is considered for the study as shown in Figure 13.1. The mechanism architecture is discussed using the kinematic model. The design of the model is dependent on the fundamental mechanical elements which has been covered.

(a) (b)

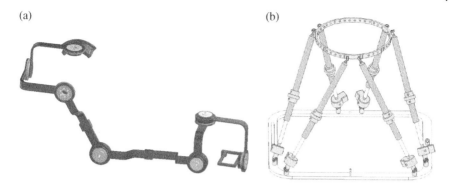

Figure 13.1 (a) Serial or open-chain arm exoskeleton, (b) parallel or closed chain Stewart platform.

13.2 Mechanisms in a Robot

13.2.1 Serial Manipulator

A serial chain arm wearable exoskeleton can control the industrial robot using hand motion. The task of a peg-in-hole insertion requires the usage of an industrial manipulator (KUKA KR5 ARC). For effective functioning, the slave device, in this case, the KUKA robot, amplifies the required power to reproduce the end-effector motion of the master device, in this case the exoskeleton. This kind of control known as position–position teleoperation is used. Figure 13.2 shows the arm exoskeleton worn by the operator. The anthropometric exoskeleton (shown schematically in Figure 13.2) and the human arm both have seven degrees of freedom (DOFs). The shoulder joint has three degrees of freedom, the elbow joint has two, and the wrist joint has two. The exoskeleton has sensors positioned in seven separate locations to track changes in joint angles. For the purpose of measuring the angle of rotation of abduction–adduction (inside-outside) and flexion-extension (forward-backward) motions, rotary potentiometers are utilized.

13.2.2 Parallel Manipulator

A 6-DOF parallel manipulator has a moving platform and a base platform joined by six legs. Universal joints bind the legs to the base platform, and spherical joints join the legs to the moving platform as shown in Figure 13.3. In order to regulate the movement of the moving platform, each leg contains a prismatic actuator. A Stewart platform, a general type 6-DOF parallel manipulator, was presented as a flight simulator (Stewart 1965). For the kinematic and dynamic examination of this class of manipulator, a substantial amount of research is accessible. The Stewart platform's

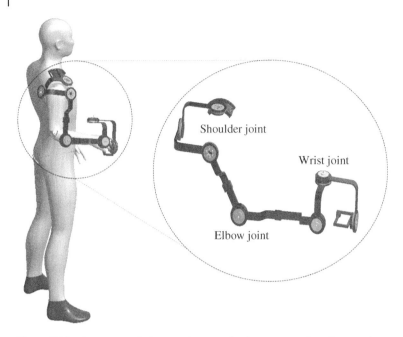

Figure 13.2 An arm exoskeleton to be worn by the operator for teleoperation.

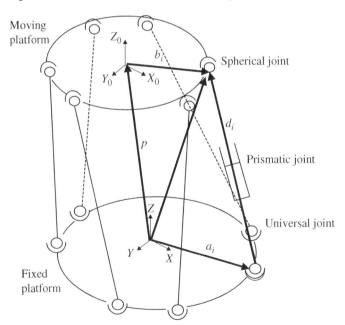

Figure 13.3 A general form of the Stewart platform. *Source:* Adapted from Tsai (1999).

forward kinematic formulation is complex and has several solutions, which is one of its drawbacks (Nanua et al. 1990; Liu et al. 1993; Innocenti 2001).

A 1-DOF parallel manipulator is often known as a Sarrus mechanism. The mechanism connects the top platform with the bottom platform with two legs. The moving platform is supported by two links on each leg, each of which is joined by a revolute joint as shown in Figure 13.4. Fre'de'ric Sarrus created it in 1853. Without the need of guide rails, it transforms rotary input to any one of the six joints into a spatial straight-line motion. Sarrus mechanism kinematic and dynamic studies were published in (Lee 1996; Angeles 2004; Chen et al. 2013).

Kinematic singularity resolution is one of the difficult problems in the utilization of parallel manipulators. In spite of the control action, the system may in a singularity configuration lose the ability to move in particular modes or acquire the ability to confine particular loads. The term "loss or gain of degrees of freedom" is used to describe this (DOFs). Gosselin (1990) examined the nature of singularities in hybrid systems with series–parallel combinations as well as singularities in general. Inverse kinematic singularities, forward kinematic singularities, and combinations of both types of singularities made up the three categories in which the kinematic singularities occurred. At singularities of inverse kinematics, the mechanism loses one or more degrees of freedom. Forward kinematic singularities are more difficult to forecast because the system gains one or more

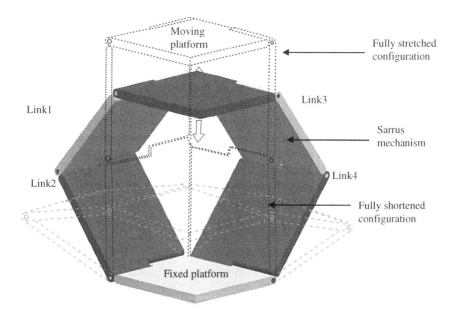

Figure 13.4 Sarrus mechanism having three different positions. *Source:* Adapted from Chen et al. (2013).

degrees of freedom, while in combined kinematic singularities, the system could, depending on the design, lose or gain DOFs. It is possible to avoid inverse kinematic singularities at the workspace limits of a mechanism by specifying an available work volume that is a subset of the geometric range. Conversely, the direct kinematic singularities are challenging to overcome because they occur on manifolds that fall inside the geometric range of parallel mechanisms. Architectural singularities, another sort of singularity that appears in 6-DOF parallel manipulators, are also described by Ma and Angeles (1991). This can be avoided during the design phase by selecting the manipulator's architecture appropriately.

13.3 Kinematics

Kinematics is defined as the relationship between joint motion and the end-effector motions without considering the causes of the motions. Kinematics is broadly classified into two problems, that of direct kinematics and inverse kinematics (Craig 2005). Forward kinematics is defined as the problem of determining the end-effector motion for a manipulator given the joint motions, whereas inverse kinematics is defined as the problem of determining the joint motion, for a given end-effector motion.

In a robotic system, joint motions govern end-effector movements, and it is essential to understand the relationships between joint motion (input) and end-effector motion (output) while directing a robot. In order to conduct transformations between the coordinate frames associated with the various connections of the robot, the knowledge of kinematics is crucial. A robot is composed of several connections connected serially by joints. The number of joints and links, the types of joints, and the kinematic chain of the robot all affect the DOF of the robot. Links are the individual parts that make up a robot. Unless otherwise specified, all links are presumed to be rigid in this context, which means that the distance between any two spots within the body remains constant as it moves. In three-dimensional Cartesian space, a rigid body has six DOF. This suggests that the body's orientation and position can each be defined by three rotational coordinates and three translational coordinates, respectively. Chains, cables, and belts are examples of non-rigid bodies that, when used to perform the same job as rigid bodies, can be conveniently referred to as links. When two or more members are joined together without experiencing any relative motion, this is referred to as a rigid link from a kinematic perspective.

Kinematic pairs or joints connect the links of a robot. A joint connects two links and limits the relative motion of the links physically. It is merely a concept that enables one to express how one connection moves in relation to another one; it is not an actual thing. For instance, a door's hinge joint enables rotation about an axis with respect to a fixed wall. Other motions are not feasible. The kind of relative motion that a joint allows is determined by the nature of the contact surface between the parts, which can be a surface, a line, or a point. As a result, they are referred to as lower pair joints or higher pair joints. If two mating links are in surface contact at a joint, the joint is said to be a lower pair joint. On the other hand, if the links are in line or point contact, the joint is known as a higher pair joint. The definition states that a ball rolling on a plane forms a higher pair joint than a door's hinge joint. The following list includes all potential joint types:

1) **Revolute joint**: As shown in Figure 13.5 (a), a revolute joint, sometimes referred to as a turning pair, hinge, or pin joint, allows two links to rotate relative to one another along the joint's axis. Hence, a revolute joint imposes

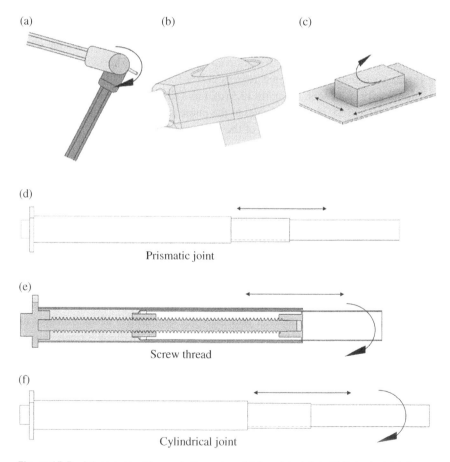

(a)   (b)   (c)

(d)

Prismatic joint

(e)

Screw thread

(f)

Cylindrical joint

Figure 13.5 Joints present in a robotic system (a) Revolute Joint, (b) Spherical Joint, (c) Planar Joint, (d) Prismatic Joint, (e) Helical Joint, and (f) Cylindrical Joint.

five limitations, i.e., it forbids rotation of two axes, as well as translation along three perpendicular axes with respect to the other link. The revolute joint has one degree of freedom (DOF).

2) **Prismatic joint**: As seen in Figure 13.5 (d), a sliding pair or prismatic junction allows two links to slide with respect to one another along its axis. Moreover, it places five constraints on it, giving it one DOF.

3) **Helical joint**: as seen in Figure 13.5 (e), enables two paired links to simultaneously rotate and translate along the joint's axis. However, the translation is dependent on the rotational motion. It is connected to rotation via the screw's pitch. As a result, the helical joint likewise has five restrictions and one DOF.

4) **Cylindrical joint**: As seen in Figure 13.5 (f), it allows independent translation motion and rotation about the joint's axis. As a result, a cylindrical joint has two DOF and places four constraints on the paired links.

5) **Spherical joint**: As seen in Figure 13.5 (b), it enables one of the connected links to freely spin about the center of a sphere in any direction feasible with regard to the other link. No translation motion is allowed. Therefore, it imposes three constraints and has three degrees of freedom.

6) **Planar joints**: As seen in Figure 13.5 (c), It is a three-DOF joint that allows for one rotation along the axis normal to the plane and two translations along the two independently moving axes of the plane of contact.

The joints are connected together to form a kinematic chain. A kinematic chain is said to be simple if each and every link in it is connected to no more than two other links. The kinematic chain can be closed or opened. All robotic manipulators fit into these groups.

13.3.1 Degree of Freedom

The formal definition of a mechanical system's degree of freedom (DOF) is the number of independent coordinates or the lowest number of coordinates necessary to fully describe its posture or configuration. A rigid body has six degrees of freedom (DOF), three for position and three for orientation when it moves in three dimensions of Cartesian space. There are various methods for calculating the DOF. Grubler offered one such technique for planar mechanisms in 1917, and Kutzbach generalized it for spatial mechanisms in 1929. They are collectively referred to as the Grubler–Kutzbach criterion, which is derived as follows:

Assuming,

d: workspace dimension.
r_m: rigid body in the robot.
c_s: total joint constraints present in the robot.
DoF: Degree of freedom of the robot

The degree of freedom of the robot can be evaluated using the following formula:

$$\mathrm{DoF} = d\,(r_\mathrm{m} - 1) - c_\mathrm{s}$$

13.3.2 Position and Orientation in a Robotic System

In the three-dimensional Cartesian space, there are two types of rigid-body motion: translation and rotation. While rotation can be defined with just three angular coordinates, translation requires three Cartesian coordinates. As a result, the rigid-body motion can be completely defined by just six coordinates. The location and orientation of various bodies in space are a continual concern while studying the kinematics of robot manipulators. Among the areas of interest are the connections between the manipulator, tools, and workpiece. A fixed-reference coordinate system, also known as the fixed frame, is built in order to define the position and orientation of a body, also known as its pose or configuration. Next, the corresponding Cartesian coordinate system is employed to characterize the attitude of the moving body.

The six independent parameters can be used to determine a rigid body's pose or its location and orientation in relation to the reference coordinate system. Let the X–Y–Z coordinate system serve as the fixed-reference frame. The U–V–W coordinate system, also known as the moving frame, is coupled to the moving body. If the position of the moving frame in relation to the stationary frame is known, it follows that the pose or configuration of the rigid body is also known.

To determine the position of the moving frame, it can be calculated using the following formulations:

$$P_{xyz} = \begin{bmatrix} 1 & 0 & 0 & tx \\ 0 & 1 & 0 & ty \\ 0 & 0 & 1 & tz \\ 0 & 0 & 0 & 1 \end{bmatrix} \tag{13.1}$$

where, P_{xyz} in equation (13.1) is the translation matrix representing translation by a length of tx and ty and tz about X, Y, and Z-axis, successively.

The orientation of the moving frame can be calculated using the following formulations:

$$R_x = \begin{bmatrix} 1 & 0 & 0 & 0 \\ 0 & \cos\theta_x & -\sin\theta_x & 0 \\ 0 & \sin\theta_x & \cos\theta_x & 0 \\ 0 & 0 & 0 & 1 \end{bmatrix}, \tag{13.2}$$

$$R_y = \begin{bmatrix} \cos\theta_y & 0 & \sin\theta_y & 0 \\ 0 & 1 & 0 & 0 \\ -\sin\theta_y & 0 & \cos\theta_y & 0 \\ 0 & 0 & 0 & 1 \end{bmatrix} \text{and} \qquad (13.3)$$

$$R_z = \begin{bmatrix} \cos\theta_z & -\sin\theta_z & 0 & 0 \\ \sin\theta_z & \cos\theta_z & 0 & 0 \\ 0 & 0 & 1 & 0 \\ 0 & 0 & 0 & 1 \end{bmatrix} \qquad (13.4)$$

where, R_x, R_y, and R_z in equations (13.2), (13.3), and (13.4) are the rotation matrices representing rotations by angles θ_x, θ_y and θ_z about X, Y, and Z-axis, successively.

There are various ways to explain how a rigid body is oriented in relation to the fixed frame.

1) Direction cosine representation
2) Rotations on fixed axes
3) The depiction of Euler angles
4) Rotations on single and multiple axes
5) Euler parameters, among other things.

Each one comes with its own restrictions. To get around the constraints of the first representation, one can, if necessary, transition from one to the other while a robot is moving. This is an explanation of these representations.

13.4 Conclusion

Robotics is becoming increasingly important in engineering education as it provides students with the skills and knowledge necessary to design, build, and program robots that can perform a wide range of tasks in different environments. However, it is important to recognize that robots are not a replacement for human workers, but rather a tool to assist and enhance human capabilities. Robots are particularly useful for performing tasks that are dangerous, repetitive, or require high precision, as they can work tirelessly without getting tired or making mistakes. This can improve productivity and efficiency, especially for tasks that require large-scale production or continuous operation. It is crucial to have a clear understanding of the benefits and limitations of robotics before placing them into a particular application or environment. This includes considering factors such as cost-effectiveness, safety, and the potential impact on human workers. Overall, the integration of robotics into engineering education provides students with valuable skills and knowledge that can be applied to a wide range of industries and applications. However, it is important to approach robotics with a clear understanding of its capabilities and limitations and to use it in ways that enhance human capabilities rather than replace them.

Acknowledgment

Adamas University, Kolkata, India, Indian Institute of Technology Delhi, New Delhi, India, Chungnam National University, Daejeon, South Korea owe thanks for the support and excellent research facilities.

Conflict of Interest

The author affirms that they have no known financial conflicts of interest or close personal ties that would have appeared to have an impact on the work described in this chapter.

References

Angeles, J. (2004). The qualitative synthesis of parallel manipulators. *Journal of Mechanical Design* 126: 617–624.

Chen, G., Zhang, S., and Li, G. (2013). Multistable behaviors of compliant sarrus mechanisms. *Journal of Mechanisms and Robotics* 5: 021005.

Craig, J.J. (2005). *Introduction to Robotics: Mechanics and Control.* Upper Saddle River, NJ: Pearson/Prentice Hall.

Floreano, D. and Wood, R.J. (2015). Science, technology and the future of small autonomous drones. *Nature* 521 (7553): 460–466.

Fukuda, T., Kawamoto, A., Arai, F., and Matsuura, H. (1994). Mechanism and swimming experiment of micro mobile robot in water. *Proceedings of the 1994 IEEE International Conference on Robotics and Automation* (8–13 May 1994), pp. 814–819. IEEE.

Gosselin, C. (1990). Determination of the workspace of 6-dof parallel manipulators. *American Society of Mechanical Engineers* 112: 331–336.

Helgeson, M.D., Lehman, R.A., Sasso, R.C. et al. (2011). Biomechanical analysis of occipitocervical stability afforded by three fixation techniques. *The Spine Journal* 11: 245–250.

Hudson, J., Orviska, M., and Hunady, J. (2019). People's attitudes to autonomous vehicles. *Transportation Research Part A: Policy and Practice* 121: 164–176.

Innocenti, C. (2001). Forward kinematics in polynomial form of the general stewart platform. *Transactions-American Society of Mechanical Engineers Journal of Mechanical Design* 123: 254–260.

Javaid, M., Haleem, A., Singh, R.P., and Suman, R. (2021). Substantial capabilities of robotics in enhancing industry 4.0 implementation. *Cognitive Robotics* 1: 58–75.

Lee, C.-C. (1996). Kinematic analysis and dimensional synthesis of general-type Sarrus mechanism. *JSME International Journal. Ser. C, Dynamics, Control, Robotics, Design and Manufacturing* 39: 790–799.

Liu, K., Fitzgerald, J.M., and Lewis, F.L. (1993). Kinematic analysis of a Stewart platform manipulator. *IEEE Transactions on Industrial Electronics* 40: 282–293.

Ma, O. and Angeles, J. (1991). Optimum architecture design of platform manipulators. *Advanced Robotics, 1991. 'Robots in Unstructured Environments', 91 ICAR., Fifth International Conference on* (19–22 June 1991), pp. 1130–1135. Vol.2. Pisa: IEEE.

Mukherjee, S., Zubair, M., Suthar, B., and Kansal, S. (2013a). Closed loop autonomous calibration of tele-operation exoskeleton. *Proceedings of Conference on Advances in Robotics* (July), pp. 1–3.

Mukherjee, S., Zubair, M., Suthar, B., and Kansal, S. (2013b). Exoskeleton for tele-operation of industrial robot. *Proceedings of Conference on Advances in Robotics* (July), pp. 1–5.

Nagarsheth, H.J., Savsani, P.V., and Patel, M.A. (2008). Modeling and dynamics of human arm. *2008 IEEE International Conference on Automation Science and Engineering* (23–26 August 2008), pp. 924–928. Arlington, VA: IEEE.

Nanua, P., Waldron, K.J., and Murthy, V. (1990). Direct kinematic solution of a Stewart platform. *IEEE Transactions on Robotics and Automation* 6: 438–444.

Ramanathan, R., Eberhardt, S.P., Rahman, T. et al. (2000). Analysis of arm trajectories of everyday tasks for the development of an upper-limb orthosis. *IEEE Transactions on Rehabilitation Engineering* 8 (1): 60–70.

Singh, B., Sellappan, N., and Kumaradhas, P. (2013). Evolution of industrial robots and their applications. *International Journal of Emerging Technology and Advanced Engineering* 3 (5): 763–768.

Stewart, D. (1965). A platform with six degrees of freedom. *Proceedings of the Institution of Mechanical Engineers* 180: 371–386.

Tsai, L.-W. (1999). *Robot Analysis: The Mechanics of Serial and Parallel Manipulators*. Wiley.

Wang, X., Yang, B., Tan, D. et al. (2020). Bioinspired footed soft robot with unidirectional all-terrain mobility. *Materials Today* 35: 42–49.

Zubair, M. and Jung, S. (2021). Mechanical joint design for gravity compensation of a robot manipulator. *2021 21st International Conference on Control, Automation and Systems (ICCAS)* (12–15 October 2021), pp. 915–918. Jeju, Korea: IEEE.

Zubair, M., Suthar, B., and Jung, S. (2022). Design and analysis of flexure mechanisms for human hand tremor compensation. *IEEE Access* 10: 36006–36017.

Index

.

Printed and bound by CPI Group (UK) Ltd, Croydon, CR0 4YY

27/10/2024

14580673-0001